# A Teachers' Guide to Physical Science

Galina V. Reid, MS
Philippe E. Tissot, PhD
Texas A&M University-Corpus Christi

**Kendall Hunt**
publishing company

Interior illustrations provided by Bruce Chaney.

**Kendall Hunt**
publishing company

www.kendallhunt.com
*Send all inquiries to*:
4050 Westmark Drive
Dubuque, IA  52004-1840

Printed in the United States of America
10  9  8  7  6  5  4  3

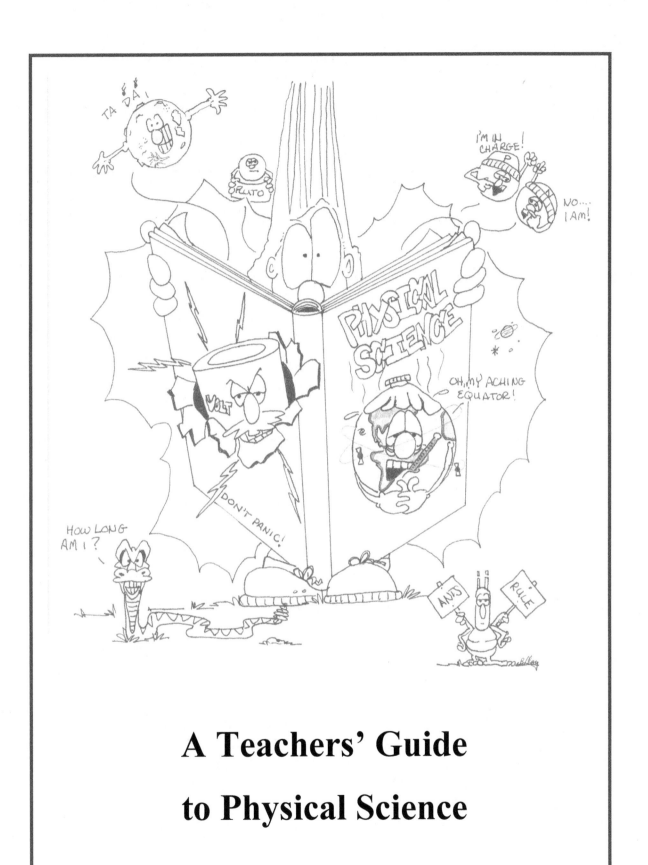

# A Teachers' Guide
# to Physical Science

# CONTENTS

| | TOPIC | PAGE |
|---|---|---|
| **1** | **Scientific Literacy and the "Scientific Method"** | **1** |
| | 1.1 Scientific Literacy | 2 |
| | 1.2 Science Teaching Methodologies and Models | 3 |
| | 1.3 Introduction to the Scientific Method | 8 |
| | 1.4 Tools of the Scientific Method: Variables | 11 |
| | 1.5 Tools of the Scientific Method: Process Skills | 14 |
| | 1.6 Tools of the Scientific Method: Measurements | 16 |
| | 1.7 Calculations | 24 |
| | 1.6 Conducting Experiments and Writing a Report | 28 |
| | 1.7 Perspective on  the Scientific Method | 36 |
| **2** | **Motion, Energy, Gravity & Simple Machines** | **37** |
| | 2.1 It' All About Motion | 38 |
| | 2.2 Gravity: "What is Keeping us on the Ground?" | 48 |
| | 2.3 Simple Machines | 57 |
| **3** | **Matter, Chemistry, Water & Heat** | **63** |
| | 3.1 It's All About Matter | 64 |
| | 3.2 Building Blocks of Matter | 67 |
| | 3.3 Where do all these Electrons Fit???! | 69 |
| | 3.4 Chemical Changes vs. Physical Changes | 75 |
| | 3.5 Pure Substances and Mixtures | 77 |
| | 3.6 Chemical Bonds | 79 |
| | 3.7 Chemical Terminology | 81 |
| | 3.8 Acidic and Basic Solutions | 82 |
| | 3.9 Water: An essential Molecule | 87 |
| | 3.10 Heat and Temperature | 95 |

**4    Electromagnetism & Waves**                                        **99**

    4.1 Electrical Charge and Charging                     100

    4.2 Electrical Polarization                           104

    4.3 Electric Current and Electrical Circuits          105

    4.4 Voltage, Amperage, and Resistance                 109

    4.5 Magnets and Magnetism                             113

    4.6 Waves Around Us                                   117

    4.7 Sound                                             120

    4.8 Light                                             122

    4.9 Reflection and Refraction of Light                124

**5    The Earth and Beyond**                                            **127**

    5.1 It's All About the Earth                          128

    5.2 The Atmosphere                                    133

    5.3 The Hydrosphere                                   138

    5.4 The Geosphere                                     142

    5.5 Global Climate Change                             152

    5.6 Beyond the Earth                                  157

# LIST OF ACTIVITIES

|  | ACTIVITIES | PAGE |
|---|---|---|
| 1.1 | Physical and Life Sciences around us | 1 |
| 1.2 | Buoyancy and the 5E Method | 6 |
| 1.3 | Investigate with the Scientific Method | 10 |
| 1.4 | Variable Identification | 14 |
| 1.5 | Practicing Metric System Conversions | 18 |
| 1.6 | Can you Handle the Metric System? | 19 |
| 1.7 | Let's Measure | 21 |
| 1.8 | Measure and Compute | 26 |
| 1.9 | Pendulum Experiments | 29 |
| 1.10 | Little Car Experiments | 32 |
| 2.1 | Observing Motion – Bumper Cups | 40 |
| 2.2 | A Force of Attraction | 50 |
| 2.3 | How Do Objects Fall to the Ground | 52 |
| 2.4 | Modeling the Motion of a Rocket | 54 |
| 2.5 | Lift it up! | 59 |
| 3.1 | Decode the Periodic Table | 68 |
| 3.2 | Decode the Electronic Configuration | 71 |
| 3.3 | Decode the Molecules | 73 |
| 3.4 | Growing Crystals | 76 |
| 3.5 | Mixtures | 78 |
| 3.6 | Sort it all out | 78 |
| 3.7 | Acidic or Basic? Can Color Tell? | 83 |
| 3.8 | Water: the 'Mickey Mouse' Molecule? | 88 |
| 3.9 | Is Water Sticky? | 89 |
| 3.10 | How many Drops of Water Will Fit on a Penny? | 90 |
| 3.11 | How Substances Dissolve | 92 |
| 3.12 | Float or Sink? It is a Matter of Buyancy! | 93 |
| 3.13 | Ice Cream | 97 |
| 4.1 | Electrical Charge and Charging | 102 |
| 4.2 | Electrical Current | 106 |

| | | |
|---|---|---|
| 4.3 | Parallel and Series Circuits | 110 |
| 4.4 | How to Make a Circuit Maze | 111 |
| 4.5 | Fishing Expedition! | 114 |
| 4.6 | Fishing Expedition II | 116 |
| 4.7 | Standing Waves | 119 |
| 4.8 | Example of Activity Designed for Special Education | 121 |
| 4.9 | Sound vs. Light | 123 |
| 4.10 | Mirrors, Lenses and Prisms | 125 |
| 5.1 | An Interacting System | 132 |
| 5.2 | The Imploding Soda Can | 134 |
| 5.3 | Good Ozone – Bad Ozone (web activity) | 136 |
| 5.4 | Where is the Water? | 138 |
| 5.5 | Water Drainage | 140 |
| 5.6 | Currents and Climate (web activity) | 141 |
| 5.7 | The Earth and a Boiled Egg? | 142 |
| 5.8 | What's Beneath our Feet? | 143 |
| 5.9 | Tectonic Plates | 147 |
| 5.10 | Plate Tectonics and Geological Activity | 148 |
| 5.11 | Rock Cycle Dance | 151 |
| 5.12 | Graphing Climate Change (web activity) | 153 |
| 5.13 | Solar Energy Accounting | 155 |
| 5.14 | Phases of the Moon | 163 |

# Preface

This book started as a set of class notes that we began assembling in the spring of 2004. The class notes were originally based on a package of activities that our predecessor Dr. Robert MacDonald had skillfully used in his classes. His class notes followed a long tradition of hands-on learning for future educators at our university established by Drs. Joyce and Janice Freeman. The goal of this book is to guide future teachers in their study of Physical Science through a set of activities and charts/diagrams. The activities are designed such that they can be easily adapted to various educational needs. The goal of the charts and diagrams is to structure and visualize the connections between the main concepts of broad topics. We hope to have created a good mix of adapted classics and new activities along with the concentrated content presentation. This book is designed to be a guide during the learning process and a quick reference but by no means an exhaustive source on the topic. We advise our students and readers to acquire at least one more complete book on Physical Science to accompany this guide. We hope you will enjoy this guide, get enthusiastic about teaching Physical Science and let us know of what works and what does not. This book would not have been possible without the feedback of our former students, teachers from the local school districts, and long discussions with our colleagues Nancy Darnell and Dr. Cherie McCollough. We thank all of them for their input, comments, suggestions, and encouragements to continue building and improving these class notes.

Galina Reid and Philippe Tissot, May 27, 2009

# Topic 1

# Scientific Literacy and "The Scientific Method"

# 1.1 Scientific Literacy

Throughout this guide we will learn and apply scientific facts, the related theories and how knowledge is gained through the scientific method. We will do so in part by reviewing and discussing scientific content and by be performing experiments and developing our understanding of the natural world through these experiments.

This guide's approach to scientific literacy and the teaching of science is to combine the guidance to acquire content while including opportunities to develop and practice the skills and techniques used by scientists (Fig. 1.1). A positive attitude towards science is very important for the development of scientific literacy and the development of such attitude should be one of the objectives of a science educator. After all Science can be quite a bit of fun so why not share it!

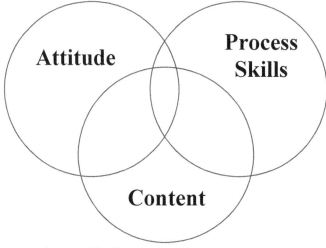

Figure 1.1. Components of scientific literacy.

As the title indicates the focus of this book is on Physical Sciences. The division between Physical and Life Sciences can be a bit arbitrary but it helps break down the content in more manageable portions. Physical Sciences will be further divided in smaller topics in this book. The main topics generally covered in Physical and Life classes are listed in the table below.

Table 1.1 Typical physical and life sciences topics.

| Physical Sciences Topics | Life Sciences Topics |
|---|---|
| Earth-Science, Motion, Forces, Energy, Sound, Light, Electricity, Magnetism, Heat, Weather, Weather, Solar System | Human Body Systems, Cells, Heredity, Plants, Animals, Ecology |

## *Activity 1.1*: "Physical and Life Sciences around us"

Go outside around your house/school, and make a list of things you can observe. Once you have about 20 items on your list place them in either the Life science or the Physical Science Category.

# 1.2 Science Teaching Methodologies and Models

In this section we introduce two generally recognized strategies to teach science. We focus on inquiry-based learning and the "5-E" model. Their application has been encouraged by associations such as the National Science Teacher Association (NSTA) and the National Science Foundation (NSF). These methods are based on the educational theory of constructivism or individuals building progressively their own understanding. Constructivism is originally based on the works of Piaget and Vygotsky. The basic idea is that students come to class with a set of prior experiences and beliefs, and they construct knowledge based on those parameters. Constructivist teachers need to account for each student's experiences, misconceptions and needs. They plan the learning experience accordingly. There are a number of other models such as control theory, brain-based learning, behaviorism, the social cognition learning model, communities of practice, observational learning or social learning theory and others. All these models are not mutually exclusive. We encourage our students to research these other methods/philosophies of instructions but refer them to methodology courses and other courses from Colleges of Education for more in-depth studies.

On June 15, 2004 NSTA released a draft position statement on Scientific Inquiry in which it ***"recommends that all K-16 teachers embrace scientific inquiry and is committed to helping educators make it the centerpiece of the science classroom".***

The Inquiry-based method is based on the scientific method itself. First a scientist comes up with a problem based upon either prior knowledge or observation; the scientist then formulates a hypothesis that will be tested in the experiment. Similarly when a child or student has a question he or she can conduct an experimental investigation to better understand the underlying concept and try to find an answer.

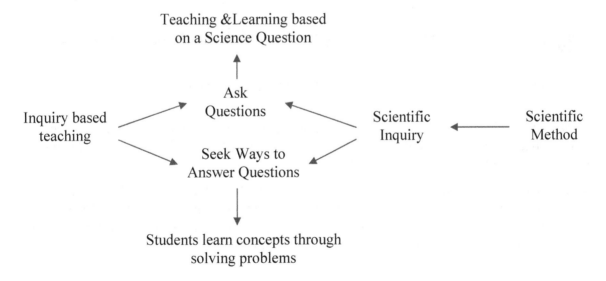

Figure 1.2. Relationship between the scientific method and inquiry based teaching.

The inquiry method implicitly teaches the scientific method and also takes advantage that children are natural scientists. They explore the natural world around them from an early age. It is natural for a child to ask a question and look for the answer in direct interaction with the environment. A student actively involved into seeking the answer obtains the content knowledge in the process and this knowledge comes through a deeper understanding then just content memorization.

A great deal of planning is necessary for the inquiry based lesson. A teacher needs to think through carefully that ladder which step by step will bring students from the original level of knowledge and understanding to one the teacher is planning to achieve. And for the inquiry based lessons that ladder must be built upon the experiment which can be a simple "hands on" activity or a long term research project. In all cases the role of the teacher is to set the stage for the inquiry and let the students to develop their conceptual understanding.

 Keep in mind that "hands-on" activities directed by a teacher should not be called "inquiry-based" unless is they are structured in such way that the students come up with both: the question and the strategy to find the answer.

In "Inquiry and the national science education standards: A guide for teaching and learning." (National Research Council (2000), Washington, DC: National Academy Press)
NSTA points scientific inquiry as "a powerful way of understanding science content". Further NSTA recommendations are illustrated in the schematic chart 1.1.


The 5- model is another recognized methodology to teach science concepts. The method emphasizes that students should be guided and encouraged to explore diverse representations of a common underlying reality and build their own understanding (with some troubleshooting and assessment from the teacher). The chronology of the 5-E model is illustrated in figure below.


# Engage → Explore → Explain → Elaborate → Evaluate


Figure 1.3. The 5-Es chronology approach to teaching a science concept.


In the Problem-based approach the students are presented with a poorly defined and complex problem that has no correct answer. The students then usually work in teams to understand the problem and work towards a solution.

We present below an example of application of the 5 methodology that can be conducted with 4th graders to explore the concept of buoyancy. The example addresses Science TEKS (4. 7. B) "conduct tests, compare data, and draw conclusions about physical properties of matter including states of matter, conduction, density, and buoyancy".

# Chart 1.1: "NSTA Recommendations and Inquiry"

**Develop short- and long-term goals**
that incorporate appropriate content knowledge.

**Bring explorations into the classroom.**
Start teaching science with explorations.
Raise and answer questions about the
natural world from those experiences.

**Select teaching strategies**
that will gradually develop
student's understandings
and abilities.

**Help students to understand:**

Science is about asking questions about the natural world and trying to answer them through the experimental investigations. There are different ways to conduct scientific investigations depending on the problem but they all are based on collecting evidence. Throughout history our perception of the world has changed due to scientific investigations. They should always be skeptical when assessing their own work or other's work. To learn about science one must understand and practice scientific inquires.

**Help students to learn how to:**

- Ask questions that can be answered through scientific investigations.
- Design and conduct investigations with use of appropriate equipment and tools to collect the evidence needed to answer the questions.
- Interpret and analyze data thinking critically about the experiment.
- Logically draw conclusions based on the evidence.
- Communicate and defend their results to their peers and others.

**Receive administrative support for the pursuit of science as inquiry in the classroom.**
- professional development on how to teach scientific inquiry.
- the allocation of time to do scientific inquiry effectively
- the availability of necessary materials and equipment.

**Experience science as inquiry as a part of teacher preparation program**
- learn how to develop questioning strategies
- learn how to write lesson plans that incorporate scientific inquiry
- learn how to analyze instructional materials to determine whether they promote scientific inquiry.

# *Activity 1.2:* **"Buoyancy and the 5E Method"**

## 1E : Engage

Objective: Stimulate curiosity and activate prior student knowledge.
 Choose an activity that will raise questions and motivate students to discover more about the concept. It could be a problem or an event that students have prior knowledge of. Remember: sometimes prior knowledge is based on limited experiences and sometimes on misconceptions.

Demonstrate to your class that a rock will sink if you drop it in a glass of water but a piece of wood will float.
Ask students "Why?" Expected answer: a rock is heavy and a piece of wood is light.
Ask students how they could determine if an object is heavy or not? Expected answer: one must weight an object.
Ask then how an aircraft carrier can float?
Remind students about the concept of "Density"

## 2E: Explore:

Objective: Actively explore the concept in a hands-on activity.
This establishes a commonly shared classroom experience and allows students to share ideas about the concept.
**Experiences must occur before the explanations!** Students learn not from a teacher but from self learning. Acquired common set of concrete experiences allows students to help each other understand the concept through social interaction.

Offer to students to determine the density of rock and wood samples using scales, graduate cylinders and rulers.
Write down on the board the results from all "research groups" and average it. Result: the density of rock is greater than the density of wood.
Give the students the density of an orange or an apple (which is less than the density of a rock but more than the density of wood) and ask to predict if the fruit will float.
Students realize that there is not enough information. They need to compare densities not to each other but to the density of water. Give the density of water to students.
Ask students to write down all the densities and compare the density of rock, wood, and fruit to the density of water. Then, based on the result of the comparison and the information about the buoyancy of rock and wood, students make a prediction about the buoyancy of a fruit and check the prediction.

# 3E: Explain

Objective: To introduce and explain new scientific terms through students' discussion of information discovered during the Explore stage.

The teacher needs to develop a questioning strategy to involve students in meaningful discussions with other students and the teacher, so they can pool their explanations based on observations, construct new understandings, and have a clear focus for additional learning.

Questions:

1. How to predict if an object will sink or float?

2. How can density tell you if an object will sink or float?

3. What does it mean "a very buoyant object"?

4. Why do object can have great buoyancy?

5. If the density of iron is greater than the density of water why do metal boats not sink?

Explain overall density of a metal boat or ship.

# 4E: Elaborate

Objective: To apply, extend, and enhance the new concept and related terms during interaction with the teacher and other students. Immediate use of new information and concept allows students to confirm and expand their understanding.

Students construct boats from aluminum foil and load them with "cargo" gradually increasing the weight of the cargo.

Goal: determine the maximum weight the boat can handle.

For an advance assignment students may calculate the overall density of aluminum boat and predict the cargo limit.

# 5E: Evaluate:

Objective: To demonstrate how students understand the concept. It is important for students to be aware of their own progress as an outcome of instruction.

Students are given a set of labeled film containers partially filled with sand and known volume. Students need to measure the mass of each container, calculate the density, and based on density predict if the container will float or sink.

# 1.3 Introduction to the Scientific Method

When studying the sciences it is important to not only learn content but also how to do science. After all this is how the content was developed. Scientists follow a methodology based on developing theories/hypothesis and testing them with experiments. The experimental discovery, or verification, is at the heart of the method. In science, for a theory to be considered "good" or "valid" it has to be verified experimentally. If it is not verified experimentally it is not part of science. Note that at times theories are ahead of our experimental abilities. In such cases the theory is not necessarily a bad one but it will have to wait for experimental verification before being truly part of our scientific knowledge. Before being considered valid, an experiment must be reviewed and approved by other scientists who have expertise in the work. This review process by other specialists is called "peer review" and is another important part of this process of discovery. This methodology to discover and understand the natural world is called the **scientific method**.

More generally the **scientific method** can be described as the way scientists investigate the world and produce knowledge about it. The term is typically used to describe a step by step systematic approach. While we will describe and practice this approach in this guide, keep in mind that many scholars do not believe that the scientific method can be described as a series of steps. They argue that the actual work of scientists is less structured. At the beginning of an investigation there is usually a good deal of preliminary unstructured investigations and brainstorming leading to a more formal scientific inquiry. In all cases the main differences between the scientific method and other methods to discover knowledge are the use of **controlled experiments** with the requirements that the results be **reproducible** and formally reviewed by other scientists.

While the discovery of knowledge about our world has been intimately linked with humanity's progress - from the early discovery of how to start a fire to forging metal objects - the development of the scientific method came much later. One usually credits the earliest foundations of the scientific method to Roger Bacon (1214-1294) in England and Galileo Galilei (1564-142) in Italy. These contributions were followed by other important ones from Francis Bacon (1561-1626), Rene Descartes (1596-1650) and others who all added to the development of the scientific method. However we believe that this set of practices which became the scientific method were inspired by earlier scientists in the Islamic world or even before.

The scientific method has been very successful at helping us discover and better understand the natural world and is intimately linked with humanity's recent history. History has shown that just using common sense or even logic can lead to fundamental misconceptions. In activity 1.3 we will work through such example that was corrected thanks to the sound application of the scientific method. There are other approaches to discoveries and understanding such as discoveries based on intuition, inspiration, pure logic without experimental verification, consensus opinion, etc. While the scientific method is clearly successful when applied to the study of our natural world, the method has limitations when applied to other fields of human endeavors. For example it is not recommended as a major tool in the arts or in fashion design.

The scientific method is the accepted method for the study of the natural world and a better understanding of the method leads to a better understanding and use of its results.

# Chart 1.2: "Steps Used in the Scientific Method"

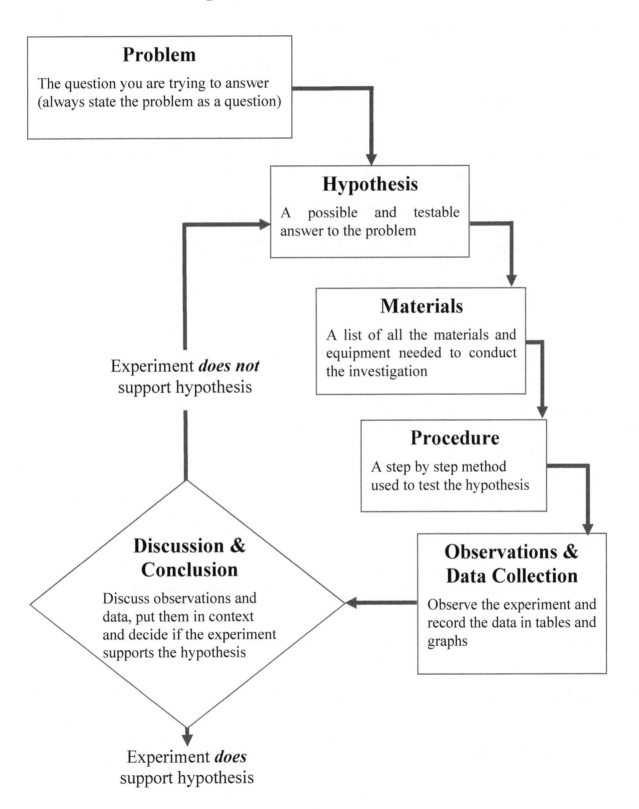

**Problem**

The question you are trying to answer (always state the problem as a question)

**Hypothesis**

A possible and testable answer to the problem

**Materials**

A list of all the materials and equipment needed to conduct the investigation

**Procedure**

A step by step method used to test the hypothesis

**Observations & Data Collection**

Observe the experiment and record the data in tables and graphs

**Discussion & Conclusion**

Discuss observations and data, put them in context and decide if the experiment supports the hypothesis

Experiment *does not* support hypothesis

Experiment *does* support hypothesis

## *Activity 1.3:* **"Investigate with the Scientific Method"**

We will now apply to the scientific method to answer a question:

**Problem: "How do objects fall?"**

Following the scientific method we will test hypothesis and set-up an experiment to test two hypotheses.

The two hypotheses come from two famous scientists and philosopher, the Greek Aristotle (384-322 BC) and the Italian Galileo Galilei (1564-1642).

**Aristotle Hypothesis: "Heavier objects fall more quickly then lighter ones"**

**Galileo Hypothesis: "Objects of different weights will fall at the same speed"**

Your hypothesis about which one of them is right: _____

**Materials:** (one set for each scientific research team)
- Set of marbles that are same in size but different in weight
- Stopwatch

**Procedure:** (each team member should have a specific job)
- Set up an experiment where marbles are released one by one from the same height.
- Time the fall of the each marble from the set height and record the results into a table.
- Repeat each experiment three times to assure accuracy then average it for each marble.

**Discussion/Conclusion:**
- Look at average time for each marble. How large are the differences in time? Compare it with the differences in weight. Based on the experimental results discuss the hypothesis and conclude if you can

Is your conclusion different from your hypothesis?


How the result of this experiment correlates with your life experience?


If there is a contrivers discuss "why?"

# 1.4 Tools of the Scientific Method: Variables

An essential part of the scientific method is to check if a hypothesis is correct or not correct by conducting an experiment. In this section we explore the tools that are essential for scientific experimentation. It is important for the person conducting the experiment to have some process skills: to understand how to select variables, to know how estimate the precision and accuracy of the measurements, to be able to interpret the experimental results. Also most scientific experiments use the metric system. We will study the logic of Metric System and learn how to convert units in the metric system, and then practice the conversions.

At the start of a scientific experiment one needs to form a strategy:

*How to set up the experiment? What should we vary/not vary? What should we measure? ...*

Scientists try to keep things as simple as possible and change as few parameters as possible between measurements. Ideally one tries to change only one parameter (variable) at a time. A variable is a measurable quantity which value can be changed. There are three types of variables as illustrated in the chart below.

## Chart 1.3:    "Variables in a Scientific Experiment"

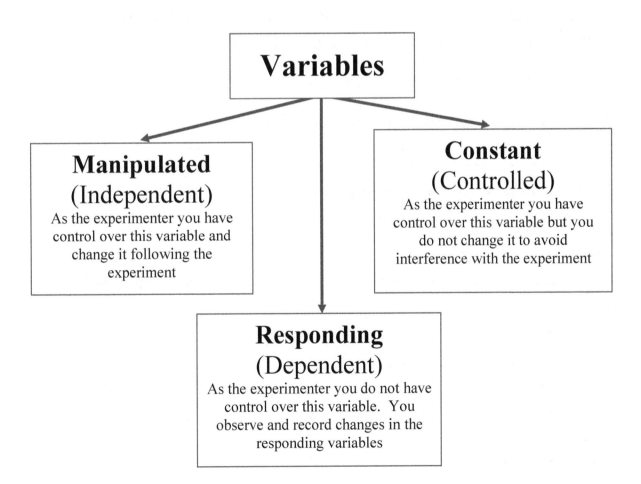

Let' start with an example to better understand the three types of variables:

*A student is testing how the temperature of tea affects the melting time of an ice cube.*

What is **the cause** and what is **the effect**? The tea is causing the ice cube to melt. Therefore the time of melting (a parameter that belongs to the ice cube) will respond to the temperature of the tea (a parameter belongs to the tea). This tells you which parameter will be dependent (responding) variable in this experiment; it is the time it takes to melt an ice cube. And the parameter causing the response will be independent (manipulated) variable. In this example it is the temperature of the tea. But there are other factors that could influence the time of melting such as amount of tea in a cap. The effect of these factors must be minimized to clear the result of the experiment. What parameters need to be taken under control in the above experiment? As it has already been mentioned the amount of tea in a cap, than the size of an ice cube. Another factor that could influence the results is the heat capacity of the cap. All this parameters will make controlled or constant variables. The experimenter should try to keep them the same throughout the experiment.

So, there is the summary:

Independent Variable – Temperature of the tea.

Dependent Variable – Time of melting.

Controlled Variables – Amount of tea, size of a cube and heat capacity of the cap.

**Exercise:** *A student wants to test how texture of the surface affects the distance a toy car goes.*

1) What parameters should this student change during the experiment?

_____

_____

2) What parameters should this student keep from changing?

_____

_____

3) Name variables of each type for this experiment.

_____

_____

_____

4) *A student wants to test how porosity of the soil affects the time water drains through.*

Manipulated Variable?_____

Responding Variable?_____

Constant Variable?_____

5) *A student wants to test if the brightness of a bulb will be affected by the voltage of a battery.*

Manipulated Variable?_____

Responding Variable?_____

Constant Variable?_____

## Activity 1.4:   "Variable Identification"

*A. Test how the temperature of tea affects the melting time of an ice cube.*

*B. Test how the texture of a surface affects the distance a toy car goes.*

*C. Test how the  porosity of soils affects the time water drains through.*

*D. Test if the brightness of a bulb will be affected by the voltage of a battery.*

Each of the four experiments mentioned above is simple enough to set in a classroom so students of K-3 grades can learn how to deal with variables in a science experiment.

- For each example find an appropriate TEKS that could be addressed in grades K-3.

- For each example state the teaching goal you could reach.

- Choose one of the examples and write down a structural plan that could be incorporated into any lesson plan. Carefully think about each step of your plan: what you are trying to achieve and how effective would be the assessment.

|   | Grade | TEKS | Teaching Goal |
|---|-------|------|---------------|
| **A** |   |   |   |
| **B** |   |   |   |
| **C** |   |   |   |
| **D** |   |   |   |

**STRUCTURAL PLAN:**

# 1.5 Tools of the Scientific Method: Process Skills

While doing science students will use a number of process skills essential for science and also very useful in life in general. Process skills are also emphasized in science standards for elementary grades. Examples of process skills are presented in the table below.

Table 1.2 Examples of process skills in Science.

| Skill | Description | Activities |
|-------|-------------|------------|
| Observing | Using the senses to obtain information | Describing properties of objects, systems, etc. |
| Describing | Isolating important characteristics using appropriate terminology | Describing objects, shapes, events |
| Classifying | Grouping objects according to predetermined set of properties | Comparing and contrasting characteristics, sorting objects based on likeness and differences |
| Predicting | Anticipating future events based upon past observations and experiences. | Predicting how long a candle will burn under various sizes of containers. |
| Formulating Hypothesis | Process of developing an if…then statement which will then be tested by conducting an experiment | Formulate a statement which can then be verified by an experiment. |
| Measuring | Developing appropriate units of measurements for length, area, volume, time, weight, etc. | Measuring distances with spans of string, covering an area with books, counting seeds that fill a container |
| Experimenting | Verifying a hypothesis through the use of materials and the control of variables | Set up relevant experiments to help verify a hypothesis. |
| Controlling variables | Keeping all factors constant except the single factor being tested through an experiment | Identifying variables already controlled and those which may be manipulated in an experiment |
| Communicating | Compiling information in graphic or pictorial form describing objects and events in detail | Making and interpreting information from graphs, charts, maps, etc. Logically arranging data sequencing ideas and events |
| Analyzing Data | Analyzing data, evaluating information, establishing relationships and applying information to other situations | From all results of an experiment conclusions are made or left open. |
| Inferring | Interpreting direct observations. Deducting information based on the observations. Past experiences are often used as a basis. | Inferring that the moisture that collects upon a glass come from the air, inferring the characteristics of an animal from his tracks. |

**Exercise:** *A kindergarten teacher asks students to collect dry leaves, pebbles, and twigs around the play ground. Then she sorts out the collected into three groups by type and seals it into three type of tin cans marked "A", "B", and "C". During the class the teacher asks the students to determine what kind of object is sealed in each type of can without opening the can. The students shake the cans and describe the sounds. Then they discuss what cold make this sounds and why. Afterward the students guess the contents of each type of can. At the end the teacher opens each can to check if student's guess is correct.*

A) What TEKS was the teacher addressing in this activity?

_____

_____

B) Name process skills the teacher was mastering with this activity?

_____

_____

C) How this activity could be integrated with Language art TEKS?

_____

_____

D) Design your own activity that will address "Process Skills" TEKS for 1$^{st}$ and 2$^{nd}$ grade.

| Grade | 1 | 2 |
|---|---|---|
| TEKS addressed | | |
| Process skills mastered | | |
| Description of activity | | |

# 1.6 Tools of the Scientific Method: Measurements

A key process skill for science experiments is **measuring**. Let's think about what measuring means and let's practice. But before we get to the experiment let's start with the story of four friends basking under the hot Texas sun.

*A story*

Once upon a time in a Texas ranch lived a Rattle Snake, a Raccoon, a Donkey and a Sparrow. One lazy afternoon, while they were stretching after a siesta, the Raccoon said "Snake! You are so long!" And the Rattle Snake interested answered "Hmm, how long am I?" "I don't know!" replied the Raccoon.

They looked at each other puzzled. How long is the snake? They decided to ask the Sparrow.

The sparrow started counting: "Let's see! One, Two. Three..." as he stepped heel to toe along the Snake.

Finally the sparrow gave his measurement: "The snake is 98 Sparrows long!"

"Wow" exclaimed everybody.

"Can we measure him in Raccoons?" inquired the Raccoon.

"Why not? Let's try!"

And the Raccoon got up and started to step heel to toe along the Snake just like the Sparrow did.

"One..Two…Three…..He is 30 Raccoons long!!!"

"Can we measure him in Donkeys?" mumbled Donkey.

"Why not? Let's try!"

"One..Two.."

"He is only two Donkeys long!"

Upon hearing the last measurement the Rattle Snake coiled up and said "I don't want to be measured in Donkeys! I am longer in Sparrows!"

Mm…Is it really?

Figure 1.4. Measuring in the wild!

16

### *What does it mean to measure something?*

It means to compare with an established reference. For example, when measuring a length, one uses a ruler which is here the reference system. If you measured 10 inches it means that the length is 10 times longer then an inch. If you say that the distance to Houston is 250 miles it means that it is 250 times further than one mile.

There are different types of reference systems, one of them is the English system (feet, pounds,…) and another one is the metric or international system typically used in the sciences.

Table 1.3 Units to measure length, time, mass and volume

| Quantity | Tool | Symbol | English Units | Metric Units |
|---|---|---|---|---|
| Length | Ruler or tape measurer | L (or W or H) | feet (ft), miles (mi), yards(yd) | centimeter (cm), meters (m), kilometers (km) |
| Time | Stopwatch | T | seconds (s), minutes (min), hours (hr), | seconds (s), minutes (min), hours (hr) |
| Mass | spring scale, beam balance | M | pound (lb), once (oz) | gram (g), kilogram (kg) |
| Volume | graduated cylinder | V | Gallon (gl) | Milliliter (ml), liter (l), cubic centimeter ($cm^3$) |

## Rules for Conversion within the Metric System:

The metric system is all based on multiplying or dividing by $10^X$ (10 to the X power) to move from one unit to the other. The table below along with a few rules will help you convert measurements from one unit to the next.

Table 1.4 Orders of magnitudes for the units of the metric system

| $10^6$ | $10^3$ | $10^2$ | $10^1$ | $10^0$ | $10^{-1}$ | $10^{-2}$ | $10^{-3}$ | $10^{-6}$ | $10^{-9}$ |
|---|---|---|---|---|---|---|---|---|---|
| **M** | **K** | *H* | *D* | **Basic Unit** | *d* | **c** | **m** | **μ** | **n** |
| **Mega** | **Kilo** | *Hecto* | *Deca* | | *deci* | **centi** | **milli** | **Micro** | **Nano** |

How to use the table:

**1.** To convert a **smaller unit to a larger** unit the decimal point moves **to the left** as many decimal places as the difference between the powers of ten.

**2.** To convert from a *larger unit to a smaller* unit the decimal point moves *to the right* as many decimal places as the difference between the powers of ten.

**3.** Sometimes zeroes will have to be added in the direction where the decimal point is moved

# Activity 1.5: "Practicing Metric System Conversions"

In the tables below, fill in the empty boxes.

**Length**

| $10^3$ | $10^0$ | $10^{-1}$ | $10^{-2}$ | $10^{-3}$ | $10^{-6}$ | $10^{-9}$ |
|---|---|---|---|---|---|---|
| **K** | **Basic Unit** | *d* | **c** | **m** | **µ** | **n** |
| Kilometer | meter | decimeter | centimeter | millimeter | micrometer | nanometer |
| km | **m** | ... | cm | ... | ... | ... |
| ... | **3m** | 30dm | ... | ... | 3,000,000µm | ... |
| 0.0003km | **0.3m** | ... | 30cm | ... | ... | ... |

**Mass**

| $10^3$ | $10^0$ | $10^{-1}$ | $10^{-2}$ | $10^{-3}$ | $10^{-6}$ | $10^{-9}$ |
|---|---|---|---|---|---|---|
| **K** | **Basic Unit** | *d* | **c** | **m** | **µ** | **n** |
| Kilogram | **gram** | decigram | centigram | milligram | microgram | nanogram |
| kg | **g** | ... | cg | ... | ... | ... |
| ... | **50g** | 500dg | 5,000cg | ... | ... | ... |
| ... | **0.03g** | ... | 3cg | ... | ... | ... |

**Volume**

| $10^6$ | $10^3$ | $10^0$ | $10^{-1}$ | $10^{-3}$ | $10^{-6}$ | $10^{-9}$ |
|---|---|---|---|---|---|---|
| ... | ... | ... | ... | ... | ... | ... |
| ... | Kiloliter | **liter** | deciliter | milliliter | ... | ... |
| ML | kL | **L** | ... | ... | µL | nL |
| 0.0008ML | ... | **800L** | 8000dL | ... | ... | ... |
| ... | ... | **0.8L** | ... | ... | ... | ... |

**Length**

| ... | $10^0$ | ... | ... | ... | ... | ... |
|---|---|---|---|---|---|---|
| kilometer | **meter** | decimeter | centimeter | millimeter | micrometer | nanometer |
| ... | ... | 5.23dm | ... | 523mm | ... | ... |

**Mass**

| **M** | **K** | **Basic Unit** | *d* | **m** | **µ** | **n** |
|---|---|---|---|---|---|---|
| ... | ... | **gram** | ... | ... | ... | ... |
| ... | ... | ... | 32dg | ... | ... | ... |

# *Activity 1.6:* "Can You Handle the Metric System?"

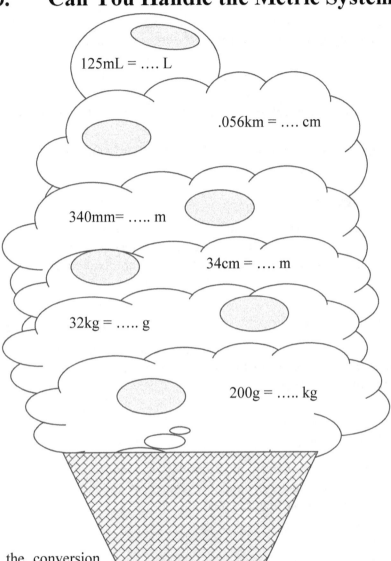

125mL = …. L

.056km = …. cm

340mm= ….. m

34cm = …. m

32kg = ….. g

200g = ….. kg

1. Complete the conversion problems stated in each of the layers of the ice-cream cone.

2. Select from the table below the letter which corresponds to the result for the conversion problems.

3. Write down the letter inside the chocolate chip.

4. If your choices are correct you will discover the hidden word!

| A | B | S | M | D | R | Y | H | U | E |
|-----|-------|-------|------|----|-------|-----|------|------|-------|
| 320 | 0.560 | 0.125 | 0.34 | 20 | 0.200 | 3.4 | 1.25 | 5600 | 32000 |

## Precision and Accuracy:

Precision and accuracy are often used as if they had the same meaning in non scientific conversations. Their meaning is however very different and the distinction is important.

*Accuracy:*    How close are the measurements/result from the true value?

***Precision:***    When an experiment is repeated how close are the measurements/results from each other?

To illustrate the difference Let's imagine that you are doing target practice with a rifle or bow and arrow.

### Accurate and Precise

The shooter hits the bull's eye every time. This is a great feat. In this case the shooting is both accurate because all the shots are close to the center of the target and precise because they are grouped, close to eachother.

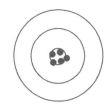

### Precise but not Accurate

All the shots are away from the center of the target and thefore the shots are not accurate. However all the shots are grouped, close to eachother and thefore they are precise.

### Accurate but not Precise

The shots are spread apart from each other and therefore not precise. On average however they are surround the center of the target. Their average would provide an accurate result.

### Neither Accurate Nor Precise

In this last case, the shots also spread apart and therfore not precise. This time if we average the location of the shots, the result is far away from the center of the target and therefore the shots are not accurate either.

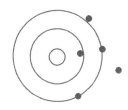

Ideally we would like to know both "how accurate" and "how precise" measurements or experiments are. To know how precise a measurement is, repeat the measurement several times and see how much difference there is between each case. Accuracy can be more difficult to assess. To know how accurate a measurement is one needs to know the "true value". This is not always possible especially if a scientist is the first to perform the measurement. One way to estimate accuracy is to perform a similar measurement for which one knows the true value.

That gives us the accuracy of the method for a similar case and this can often be a good estimate of the accuracy of the new measurement.

# *Activity 1.7:* **"Let's Measure!"**

In this activity we will measure three essential physical quantities: length, volume and mass. For each case you will discuss the measurement techniques, the meaning and limitation of the measurements, and how to present and discuss the results.

For all measurements perform the following:

- First discuss and outline a measurement strategy
- List what will affect the accuracy of the measurements
- As you test your measurement strategy estimate its precision and accuracy (i.e. if you are measuring 15 grams, could it actually be 16 grams or 17 grams or even more? or less?).
- Repeat the measurements at least three times and calculate the average
- Present the results in a table

## *Measurements of length:*

a. Measure the length of a pencil. (Individual work)

|  | Trial #1 | Trial #2 | Trial #3 | Average |
|---|---|---|---|---|
| Length, cm |  |  |  |  |

What affects the precision of the measurement?
1._____
2._____
3._____

How can you estimate the precision?_____

Estimate the precision:_____

b. Explain how you would measure the length of your partner's palm with a ruler which is broken on one end and missing some of the marks.

c. What is the average height of the students in the class? (Group work)

Divide the class in several groups, 5 or more. Each group will estimate the average height of the students in the class without measuring. Measure the height of all the students in your group and compute an average. Then measure the height of all the students in two other groups and compute their average height. After that pick a few more students in the rest of the class, measure their height and compute the average. Present the results graphically. From all of these measurements estimate the average height of the students in the class without measuring the height of all students. Discuss how many students you need to measure to get a good idea of the average student height in the class.

| | Height of the student, cm | | | | | | | |
|---|---|---|---|---|---|---|---|---|
| | #1 | #2 | #3 | #4 | #5 | #6 | #7 | average |
| Your group | | | | | | | | |
| Any other group | | | | | | | | |
| Random pick | | | | | | | | |

Can you really trust that the average will be accurate? Justify your answer.

## Measurement of volume using a graduated cylinder.

a. Measure the volume of a liquid in a plastic cup.

| | Trial #1 | Trial #2 | Trial #3 | Average |
|---|---|---|---|---|
| Volume, ml | | | | |

What affects the precision of the measurement?
1._____
2._____
3._____

How can you estimate the precision?_____

Estimate the precision:_____

b. Measure the volume of an object. To measure the volume of an object record the initial level of the water in a cylinder, then submerge the object and record the new water level. The change in levels is the volume of the object.

| | Initial water level, ml | Water level after object is submerged, ml | Volume of the object, ml |
|---|---|---|---|
| Object #1 | | | |
| Object #2 | | | |

Would this method work if the liquid was not water? Justify your answer.

_____

_____

Describe another method to determine the volume of the object?

_____

_____

## *Measurement of mass*

.   a. Measure the mass of an object by using a beam balance.
   b. Measure the mass of an object by using a spring scale

| | Trial #1 | Trial #2 | Trial #3 | Average |
|---|---|---|---|---|
| Mass using a beam balance, g | | | | |
| Mass using a spring scale, g | | | | |

What affects the precision of each measurement?

_____        _____

Estimate the precision of each measurement:

_____        _____

Which method is more precise?

What affects the accuracy of each measurement?

How could you find it out which method is more accurate?

# 1.7 Calculations

When performing scientific experiments some of variables can be measured directly. This is often the case for length, time, mass, etc. But what if you have a quantity that you can not measure directly? Then you calculate! For such variables one measures a few other quantities and then uses the measured quantities for computation. We will work below with the concepts of speed (i.e. how distance is covered per unit time) and density (i.e. mass per unit volume). The computations can also be quite useful for other purposes. If one knows the speed of a car and the distance to the destination, one can compute the time it will take to get there. If one knows the density of a material and the quantity (mass) that will be needed for a project, one can compute the volume that the material will occupy.

## Chart 1.4: "Calculations"

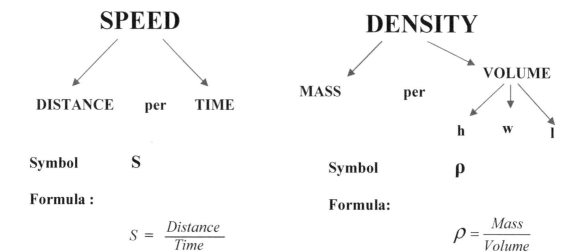

| SPEED | DENSITY |
|---|---|
| DISTANCE per TIME | MASS per VOLUME → h w l |
| Symbol $\qquad$ S | Symbol $\qquad$ $\rho$ |
| Formula : $S = \dfrac{Distance}{Time}$ | Formula: $\rho = \dfrac{Mass}{Volume}$ |
| Measurements: 1. Distance 2. Time | Measurements: 1. Mass 2. Volume |
| Units: $\dfrac{m}{s}$, mph, $\dfrac{km}{h}$, $\dfrac{?}{??}$ | Units: $\dfrac{g}{ml}$, $\dfrac{kg}{m^3}$, $\dfrac{?}{??}$ |

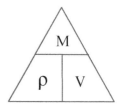

## EXAMPLES:

1. It takes for a boy 20 seconds to run 100 feet. What is his speed?

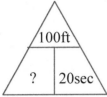

100ft : 20sec = 5 ft/sec

2. The speed of a bicycle is 15m/s. How far would it go in 10 seconds?

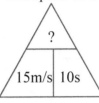

15m/s × 10s = 150 m

3. A marble of 170 g has a volume 50 cm$^3$. What is the density of material it is made of?

170g … 50cm$^3$ = …

4. The average speed of a turtle is 3cm/s. How long would it take for the turtle to cross a 180cm sidewalk?

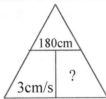

5. Mercury is a very dense liquid. Its density is 13.6g/ml. How heavy would be a glass (200ml) of mercury? Would it be reasonable to ask somebody to bring you a glass of mercury?

13.6g/ml … 200ml = …

6. The density of aluminum is 2.7 g/cm$^3$. What is the volume of an aluminum cylinder that has 10.8 grams of mass?

# *Activity 1.8:* "Measure and Compute"

In this activity you will start by measuring two quantities. Then you will use your measurements two compute a third one.

## *Computation of Fluid Densities*

Your goal is to determine the density of two different liquids. Examples of fluid that can be used for the experiment include, water, salt water and lamp oil. Use a scale and a graduated cylinder for all the necessary measurements.

Compute the density of each fluid and express the result with a proper unit.

|  | Mass, g | Volume, ml | Density,......... |
|---|---|---|---|
| Liquid #1 |  |  |  |
| Liquid#2 |  |  |  |

Are there differences in density between the two fluids and does it make sense?

## *Computation of the Density of objects*

Select a few cubes made of different materials but of similar dimensions. Your goal is to identify the materials by computing their density. You will find rulers and scales with which you can make the necessary measurements. Compute the density of each material and express the result in a proper unit.

|  | Mass, g | Volume, $cm^3$ | Density,... | Material |
|---|---|---|---|---|
| Cube #1 |  |  |  |  |
| Cube #2 |  |  |  |  |
| Cube #3 |  |  |  |  |

Are there differences in density between the materials and does it make sense?

| Material | Density | Material | Density |
|---|---|---|---|
| Lead | 11.34 $g/cm^3$ | Diamond | 3.52 $g/cm^3$ |
| Silver | 10.50 $g/cm^3$ | Aluminum | 2.70 $g/cm^3$ |
| Copper | 8.93 $g/cm^3$ | Halite | 2.16 $g/cm^3$ |
| Brass | 8.47 $g/cm^3$ | Ice | 0.92 $g/cm^3$ |
| Iron | 7.87 $g/cm^3$ | Pine Wood | 0.55 $g/cm^3$ |

## Computation of the speed of a toy car.

Gather stopwatches, measuring sticks and other measuring tools. Set-up a course and use the selected measuring tools to the speed of a toy car.  Use the Metric System.

Type of car: _____

| Trial # | Distance covered, … | Elapsed time, s | Speed,……… |
|---------|---------------------|-----------------|-----------|
|         |                     |                 |           |
|         |                     |                 |           |
|         |                     |                 |           |
|         |   Average speed: |  |  |

## Computation of the speed of a student walking, running?

Gather stopwatches, measuring sticks and other such measuring tools as needed.  Go in the corridors of a building or outside and set-up a track where you can time the students of the group walking or jogging (see what is practical and measure at least 2 speeds).  By measuring the elapsed time and the distance covered you will be able to compute the walking, jogging, or running speed.  Choose a unit of speed such as meter/second.

|         | Distance covered, m | Elapsed time, s | Speed,……… |
|---------|---------------------|-----------------|-----------|
| Walking |                     |                 |           |
| Jogging |                     |                 |           |
| Running |                     |                 |           |

Compare the speed of the toy car and the running speed of a student.

_____
_____
_____

A) How far will the toy car go in 2 minutes with the average speed determined in above?

_____
_____

B) How long would it take to the student to go across 0.2 km parking lot with the average speed determined in above?

_____
_____

# 1.8 Conducting Experiments & Writing a Report

As part of the class students will often be asked to write a formal scientific report based on experiments performed in class or individually. The report will follow the scientific method as outlined in this chapter and below. Note that there are variations to the scientific methods and you will see formal scientific reports written a little differently then recommended here but the overall structure should be similar.

The reports will start with a short title and will also include the name of the author of the report. The sections of the reports will have subtitles following the steps used in the scientific method and detailed below. If you use any references, you will finish the report with a reference section. Try to be succinct, to the point and precise in your writing.

***Problem Statement:*** A few words or a short one sentence statement describing the problem investigated (should be in question form).

***Hypothesis:*** Brief and to the point statement or statement(s) of the hypothesis tested in the experiment.

***Materials:*** A list of all the materials used in the experiment. The description should be precise enough such that another researcher having never performed the experiment and never talked to you can reproduce the experiment.

***Procedure:*** A detailed but succinct step by step description of the experiment such that another researcher can replicate the same experiment. A sketch of the experiment with labels is often a good way to help describe an experiment (there are other ways but figures/drawings are usually included).

***Observation/Data:*** Entitle this section ***Experimental Results*** and list the results of the experiment. Graphs and tables are good ways to communicate experimental results. You should not discuss the results in this section but just state the observations. Any discussion of the results should be part of the Discussion section to make a clear distinction between the observations which should be the same anywhere and the interpretation of the results which can vary depending on the scientist interpreting the results.

***Discussion/Inference:*** Discussion of the results. Analyze the main results of the experiment one by one and give your interpretation of their meaning. You can also place the results in context and compare with other similar experiments, preferably with a reference. This is your opinion of the meaning of the results, what they mean, your insight into the scientific processes involved. Different scientists can have differences of opinion when analyzing the same results.

***Conclusion:*** A few statements outlining the important points of the experiment and emphasizing what was learned or discovered.

# *Activity 1.9:* **"Pendulum Experiments"**

**Problem:**    How does the length of a pendulum effect its period?
How does the mass of a pendulum effect its period?

We just defined our problem as two questions. The next step is to make two hypotheses that we will be able to test with experiments. Write below two hypotheses or clear answers to the each question above.

**Hypothesis:** _____

_____

Now you need to think about how you will conduct this experiment, the parameters that you will measure and tools that you will use.
It is clear that *the period* will be *a responding parameter*. Think about which parameter you will keep constant and which parameter you will manipulate to have a proper response. Typically one tries to manipulate only one variable at a time.
For the first part of the experiment you could keep the mass of the pendulum constant and vary the length to see how the period responds. In this case the *mass of the pendulum* will be a *constant parameter* and *the length of the pendulum* will be a *manipulated parameter*.
For the second part of the experiment you could keep the length of the pendulum constant but vary the mass to see how the period responds.

List the materials that you are going to use in this experiment and write down the procedure separately for each part of the experiment (attach a sheet of paper). Make it as simple as possible. When writing always think about people who will read it afterward trying to repeat your experiment.

**Materials:**
I.

II.

**Procedures:**
I.

II.

## Experimental Results:

1._____ remains constant

| # of experiment | ............. | Trial #1 | Trial #2 | Trial #3 | |
|---|---|---|---|---|---|
| 1 | | | | | |
| 2 | | | | | |
| 3 | | | | | |
| 4 | | | | | |
| 5 | | | | | |

2. _____ remains constant

| # of experiment | ............... | Trial #1 | Trial #2 | Trial #3 | |
|---|---|---|---|---|---|
| 1 | | | | | |
| 2 | | | | | |
| 3 | | | | | |
| 4 | | | | | |
| 5 | | | | | |

## Data Analysis:

You can try different graphical methods to analyze the results. For example you could plot points with the manipulated variable on the horizontal axis and the responding variable on the vertical axis, and then connect them by a line. This would be a line graph that is commonly used when there is a need to illustrate how one variable is responding when another variable is manipulated. Also you could make a bar diagram using the length of bars to illustrate the values of the responding variable. A bar diagram is a good idea when the manipulated variable is not in a numerical form (different brands/ types; small-medium-large; etc.)

Your presentation must have a title and all the necessary information (axis labels, units) such that the main results of your experiment can be easily and rapidly understood directly from looking at the graph or diagram. The standard form for a graph title is: "*Responding Variable* vs. *Manipulating Variable*"

## Discussion and Conclusion (on the attached sheet of paper):

Discuss your results and make a conclusion. Go back to your hypothesis and compare the hypothesis with the experimental results. Take into account uncertainties in your comparison.

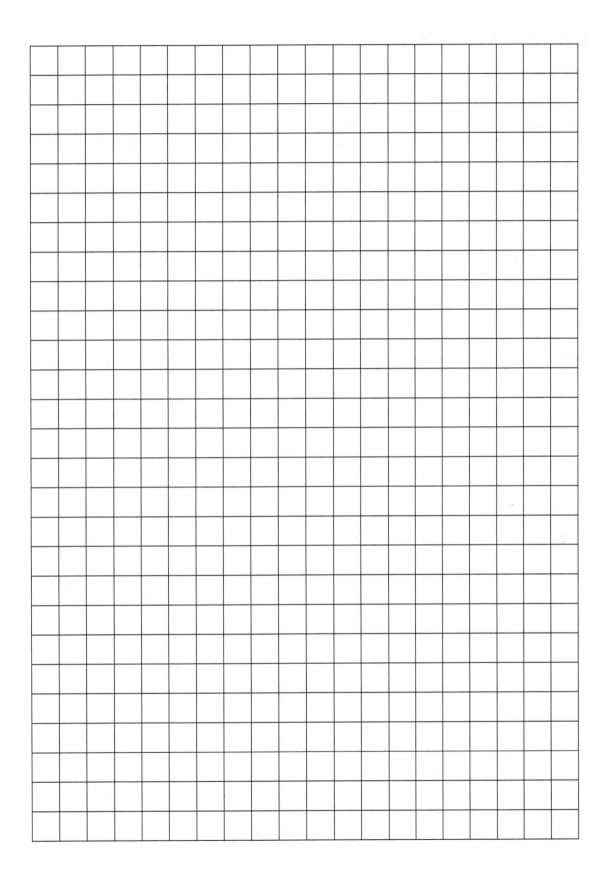

## *Activity 1.10:* "Little Car Experiments"

**Problem:**  How does the mass of a toy car affect the distance the car travels?
How does the type of surface affect the distance the car travels?

**Hypothesis:**

**Materials:**

**Procedure:**

## Experimental Results:

## Data collection:

1._____ remains constant

| # of experiment | | Trial #1 | Trial #2 | Trial #3 | |
|---|---|---|---|---|---|
| 1 | | | | | |
| 2 | | | | | |
| 3 | | | | | |
| 4 | | | | | |
| 5 | | | | | |

2. _____ remains constant

| # of experiment | | Trial #1 | Trial #2 | Trial #3 | |
|---|---|---|---|---|---|
| 1 | | | | | |
| 2 | | | | | |
| 3 | | | | | |
| 4 | | | | | |
| 5 | | | | | |

**Data Analysis:**

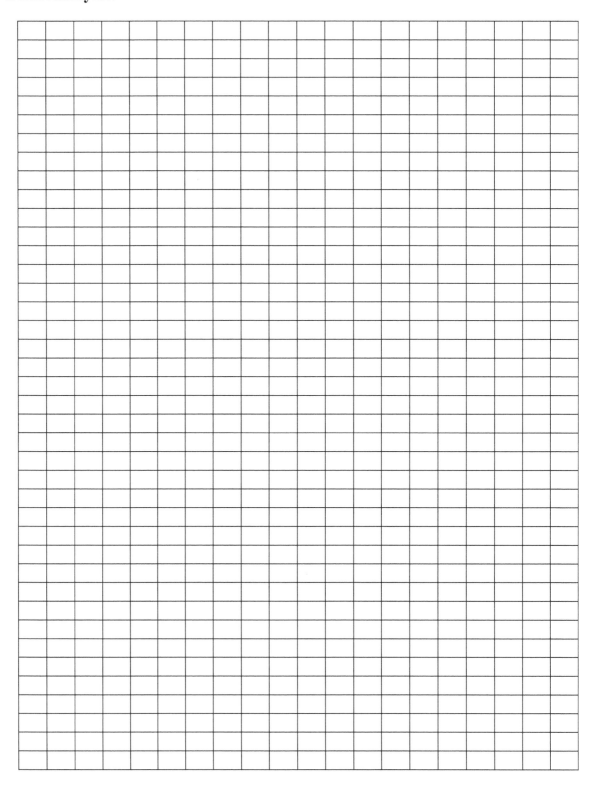

# Discussion and Conclusion:

# 1.9 Perspective on the Scientific Method

So far we presented the scientific method as a fairly structured approach to asking and answering questions about the world. Scientists do not always start and follow this exact format. At the start of a project there is usually a good deal of trial and error, "playing" with the concepts or experiments. That said scientists always keep in mind the scientific method. The closer they are from a result the more formal their approach. Pedagogically we consider teaching and practicing the formal scientific method with our students a great idea. This helps students of all ages, even the youngests, learn about science in a more concrete way and help them understand where all the scientific facts they have to learn came about. To give us some overall perspective we finish this chapter with an extract from the book Reflections of a Physicist, (1955) by Nobel Laureate Percy W. Bridgman.

"It seems to me that there is a good deal of ballyhoo about scientific method. I venture to think that the people who talk most about it are the people who do least about it. Scientific method is what working scientists do, not what other people or even they themselves may say about it. No working scientist, when he plans an experiment in the laboratory, asks himself whether he is being properly scientific, nor is he interested in whatever method he may be using *as method.* When the scientist ventures to criticize the work of his fellow scientist, as is not uncommon, he does not base his criticism on such glittering generalities as failure to follow the "scientific method," but his criticism is specific, based on some feature characteristic of the particular situation. The working scientist is always too much concerned with getting down to brass tacks to be willing to spend his time on generalities.

Scientific method is something talked about by people standing on the outside and wondering how the scientist manages to do it. These people have been able to uncover various generalities applicable to at least most of what the scientist does, but it seems to me that these generalities are not very profound, and could have been anticipated by anyone who knows enough about scientists to know what is their primary objective. I think that the objectives of all scientists have this in common--that they are all trying to get the correct answer to the particular problem in hand. This may be expressed in more pretentious language as the pursuit of truth. Now if the answer to the problem is correct there must be some way of knowing and proving that it is correct--the very meaning of truth implies the possibility of checking or verification. Hence the necessity for checking his results always inheres in what the scientist does. Furthermore, this checking must be exhaustive, for the truth of a general proposition may be disproved by a single exceptional case. A long experience has shown the scientist that various things are inimical to getting the correct answer. He has found that it is not sufficient to trust the word of his neighbor, but that if he wants to be sure, he must be able to check a result for himself. Hence the scientist is the enemy of all authoritarianism. Furthermore, he finds that he often makes mistakes himself and he must learn how to guard against them. He cannot permit himself any preconception as to what sort of results he will get, nor must he allow himself to be influenced by wishful thinking or any personal bias. All these things together give that "objectivity" to science which is often thought to be the essence of the scientific method.

But to the working scientist himself all this appears obvious and trite. What appears to him as the essence of the situation is that he is not consciously following any prescribed course of action, but feels complete freedom to utilize any method or device whatever which in the particular situation before him seems likely to yield the correct answer. In his attack on his specific problem he suffers no inhibitions of precedent or authority, but is completely free to adopt any course that his ingenuity is capable of suggesting to him. No one standing on the outside can predict what the individual scientist will do or what method he will follow. In short, science is what scientists do, and there are as many scientific methods as there are individual scientists."

# Topic 2

# Motion, Energy, Gravity & Simple Machines

# 2.1 It's All About Motion…

Whether you take a Physics or a Physical Science class, or pick a physics or physical science book it will usually start with the study of "motion". History may be one of the reasons to start with motion. After all Gallileo, Newton and their contemporary scientists developed the scientific method while figuring out how things fall, how the planets revolve around the sun, why we have tides etc. But also scientists had to start with the understanding of how things move as so many phenomena in the natural world are related to motion. Indeed Motion is all around us from the earth spinning and rotating around the sun, to the wind and the constant movement of the oceans, to cars on our roads, to birds flying and other animals moving about, to blood pumping in our veins to electrons moving to light up our houses or transfer information in our computers: it's all about motion!

Imagine sitting on a bench and having a soccer ball roll in front of you. The following questions and partial answers could come to mind:

Where is the ball coming from and how come it is still rolling in front of me?

- To answer this question you will look in the direction where the ball is coming from you know intuitively that an object in motion tends to continue in the same direction unless it bumps into an obstacle

- Implicitely you are also making the assumption that someone kicked the ball. Maybe the ball was just resting on the ground or already moving but somebody kicked it in your direction

Scientifically these observations lead to one conclusion: something acted upon the ball to set it in motion or to change its motion: we will call it a force. Forces change motion!

Newton's first law states "an object at rest will stay at rest and an object in motion will stay in the same motion unless acted upon". In other words if there is a change in motion look for the force.

Going back to our soccer ball, on earth one expects the soccer ball to eventually stop, either due to an obstacle (force) or by slowing down progressively. Before the scientific method some thought that it was a natural property of all objects to eventually stop. On earth this is generally true. However imagine kicking the soccer ball in space from the deck of a space ship far away from all planets and stars. Will the soccer ball stop? To answer go back to Newton's first law: "an object in motion stays in the same motion unless acted upon" and you will conclude that the ball will continue for ever. On earth friction and obstacles will slow down and stop the ball and mask the physical principle.

When studying motion in the following activities we will ask ourselves other questions: do all objects respond the same way to the application of a force or does it depend on a property of the object (inertia…). What happens during collisions? We will also study the first force that we experience everyday and that our scientists forefather studied first: Gravity.

# Chart 2.1: "It's All About Motion…"

**_FORCE_** - push or pull

Is a cause for a change in motion
If there is *no force applied,* an object
will have *no change in its motion*.

If you see any change in motion – LOOK FOR THE FORCE

**_INERTIA_** – is a property of an object to resist change in its motion.

The mass of an object describes how this object resists to change in motion.

The *larger* mass the *more difficult* to change the motion of an object.

To change the motion of a light object is much easier than to change the motion of a heavy one.

**_WORK_** - done by a force when changing the linear motion of an object.

**_ENERGY_** – ability of an object to do work by applying force toward another object.

An object gains energy when work is done on the object but looses energy when doing work itself on another object

 **_ENERGY_**

**Kinetic**
**Energy due to motion**
A moving object has Kinetic energy.
An object that has Kinetic energy
is moving.

**Potential**
**Energy due to position**
The further an object is
from "0" level the more
Potential energy it has

# *Activity 2.1:* "Observing Motion - Bumper Cups"

In this set of experiments we will try to answer the following questions:

- Why is the marble going down the ramp?
- What is the main factor controlling how far the marble rolls?
- What happens during the collisions between marbles and plastic cups?
- What is the main factor controlling how far the plastic cups sliding after the collisions?

To investigate these questions we will set-up a "Bumper-Cup" track and observe the motion of marbles and cups. In the first experiment we will observe the motion of marbles only. In the second experiment we will observe what happens after a marble hits a cup.

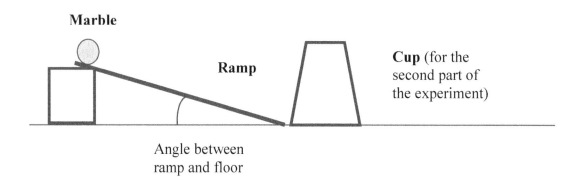

**Marble**

**Ramp**

**Cup** (for the second part of the experiment)

Angle between ramp and floor

## Experiments:

(1) a. Place a marble on the table. Does it roll? _____

b. Push the marble. Why does it roll now?

_____

c. How can you control the speed of the marble?

_____

d. Set up a ramp on the table. Place the marble on the ramp and let it go.

What do you observe? What conclusions can you make from this observation?

_____

_____

e. Let the marble roll down the ramp and time it. Vary the ramp height.

| Ramp height, cm | | | | |
|---|---|---|---|---|
| Time a marble rolls, s | | | | |

What conclusion can you make based on the data & your answer to (c) above?

_____

_____


(2) Gather a light and a heavy marble. Place the light and the heavy marble on two ramps next to each other and have them to roll down the ramps of the same height at least three times. They need to start simultaneously:  hold them by a ruler or pencil before release.

Is there a difference in the speed of the heavy and the light marbles?
How can you explain the result?

_____

_____

_____

_____


(3) Pick a *light marble* and release it from a *small angle* ramp to see how far the marble rolls. After this first experiment, predict what will happen when you increase the ramp starting height.  Vary the ramp height and let the marble roll down the ramp and onto the carpet or the table and record how far the marble rolls.  Record your results in the table and graph it.

**Prediction:**_____

| Ramp height, cm | | | | |
|---|---|---|---|---|
| Distance a marble rolls, cm | | | | |

**Marble Travel Distance vs. Ramp Height**

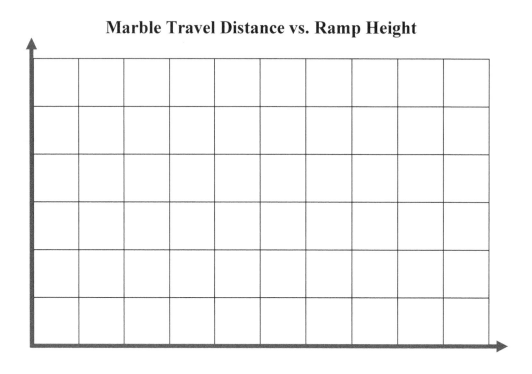

What is the main factor controlling how far the marble rolls?

_____

(4) Label what kind of energy the marble has at different points as it rolls down the ramp and slows down on the flat surface.

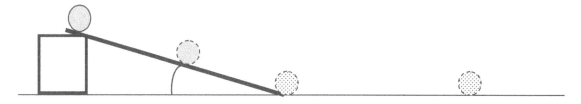

(5) Set up an experiment on a table where the light marble released from the lowest ramp hits the light plastic cup that sits on the flat surface at the bottom of the ramp. Measure the starting height of the marble with respect to the table, and how far the plastic cup is pushed after the collision.

(6) After this first experiment, predict what will happen when you :
   A) increase the ramp starting height.
      ...........................................................................
   B) increase the mass of the cup (add a play dough to the cup)
      ...........................................................................
   C) increase the mass of the marble
      ...........................................................................

42

(7) Experimentally test your predictions.

### A. The Effect of Ramp Height on the Distance a Cup Moves
*Light Cup & Light Marble - ........................ variables*

| Ramp Height, … | Distance a cup moves, …….. | | | Average Distance … |
|---|---|---|---|---|
| | Trial 1 | Trial 2 | Trial 3 | |
| | | | | |
| | | | | |
| | | | | |
| | | | | |

### B. The Effect of Cup Mass on the Distance a Cup Moves
*Light Marble & Lowest Ramp - ........................ variables*

| Mass of a Cup,… | Distance a cup moves, …….. | | | Average Distance … |
|---|---|---|---|---|
| | Trial 1 | Trial 2 | Trial 3 | |
| | | | | |
| | | | | |
| | | | | |
| | | | | |

### C. The Effect of Marble Mass on the Distance a Cup Moves
*Light cup & Lowest Ramp - ........................ variables*

| Mass of a Marble,… | Distance a cup moves, …….. | | | Average Distance,… |
|---|---|---|---|---|
| | Trial 1 | Trial 2 | Trial 3 | |
| | | | | |
| | | | | |
| | | | | |
| | | | | |

## A. Bumper Cup Travel Distance vs. Ramp Height

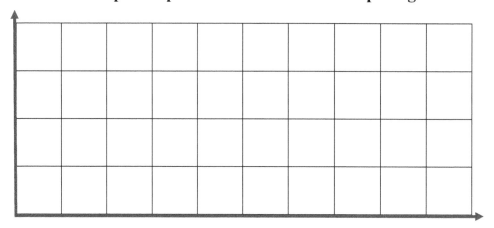

## B. Bumper Cup Travel Distance vs. Mass of Cup

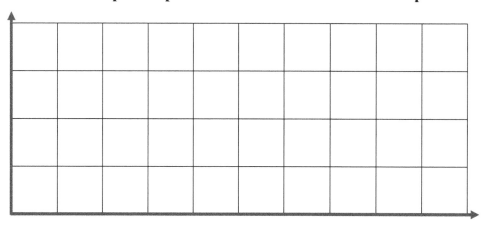

## C. Bumper Cup Travel Distance vs. Mass of Marble

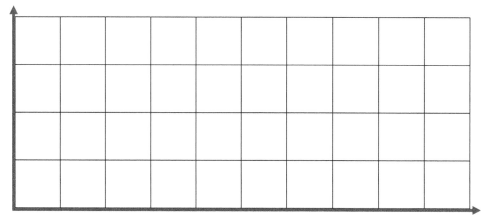

(8)  Highlight the correct part of the questions:

For a case where the masses of the cup and the marble is the same will the cup move further when the marble is released from a low or a high ramp?

Is it easier to move a light or a heavy cup?

At equal mass does a cup move more when hit by a light or a heavy marble?

State the important variables that will help you to move a cup farther.

_____

_____

_____


(9)      a. What 2$^{nd}$ grade TEKS is addressed in this activity?

_____

_____

         b. How is this activity addressing that TEKS?
            (What concepts 2 grade students will learn/what skill will gain?)

_____

_____

         c. Discuss why or why not this activity could be called
            "Inquiry based activity"

# Chart 2.2                    "Conservation of Energy"

Energy can neither be created nor destroyed, but can only converted from one form to another.

| | |
|---|---|
| I. The marble is on the highest point of the ramp and not moving.<br><br><br><br>Has it Potential Energy? .......<br>    Why would you say so? ............<br><br>Has it Kinetic Energy?    .......<br>    Why would you say so? ..............<br><br>If the ramp height increases,<br>Potential Energy ....................<br><br>If the ramp height decreases,<br>Potential Energy ....................<br>Potential Energy is ....................<br>proportional to the height of the ramp. | II. The marble is moving down the ramp.<br><br><br><br>Has it Potential Energy? .......<br>    Why would you say so? ............<br><br>Has it Kinetic Energy?    .......<br>    Why would you say so? ............<br><br>How the ratio of Potential to Kinetic energy will change as marble rolls down?<br>..............................................<br>..............................................<br>When and where this ratio equals one?<br>.............................................. |
| III. The marble is on the lowest point.<br>    The marble is still moving.<br><br><br><br>Has it Potential Energy? .......<br>    Why would you say so? ............<br><br>Has it Kinetic Energy?    .......<br>    Why would you say so? ............<br><br>Is there a relationship between the Potential Energy at the highest point and the Kinetic Energy at the lowest? Explain.<br>..............................................<br>.............................................. | IV. The marble has stopped away from the ramp. The marble is not moving.<br><br><br><br>Has it Potential Energy? .......<br>    Why would you say so? ............<br><br>Has it Kinetic Energy?    .......<br>    Why would you say so? ............<br><br>Where has energy gone?<br>..............................................<br>Why does the height of the ramp affect how far marble will go?<br>..............................................<br>.............................................. |

| V. The marble is on the lowest point and about to hit the bumper cup. | VI. The marble has hit the cup and is not moving any more. The cup is moving |
|---|---|
| | |
| Has the marble Potential Energy? | Has the marble Potential Energy? |
| …………………………………… | …………………………………… |
| Has the cup Potential Energy? | Has the cup Potential Energy? |
| …………………………………… | …………………………………… |
| Has the marble Kinetic Energy? | Has the marble Kinetic Energy? |
| …………………………………… | …………………………………… |
| Has the cup Kinetic Energy? | Has the cup Kinetic Energy? |
| …………………………………… | …………………………………… |

Newton's first law of motion states that:
**"An object continues in a state of motion at a constant speed along a straight line, unless compelled to change that state by a net force".**
Explain, based on that law, how far would the marble roll if there was no form of friction between the marble and the flat surface?

_____
_____
_____

The set of collisions between the marbles and the bumper cups illustrate another principle: *inertia.*
"Inertia is the natural tendency of an object to remain at rest or in motion at a constant speed along a straight line. The mass of an object is a quantitative measure of inertia".
Explain, based on the concept of Inertia, how does the distance the cup moves will change if you replace a light marble with a heavy marble while keeping the same cup?

_____
_____
_____

Explain, based on the concept of Inertia, how the distance the cup moves will change if you replace a light cup with a heavy cup while keeping the same marble?

_____
_____
_____
_____

# 2.2 Gravity:"What is Keeping us on the Ground?"

The unavoidable reality of Gravity is familiar to every human since childhood. "Johnny, don't climb up that chair/stair/ tree/etc., you will fall down!" cries his concerned mother and Johnny ignoring her warnings learns the rule of gravity through the painful and repeated experience. A baby dropping toys from the crib and watching them fall down is just testing a hypothesis "Do all objects fall down?" Can you imagine the excitement of the child when realizing that the hypothesis is right and ALL toys do fall down to the floor? Should the parents really be so upset about the mess on the floor?

As we grow older, we stop questioning gravity and unconsciously take it for granted in everyday life. But if gravity is something that everyone experiences every day, why did it take thousands of years to model it accurately? Isaac Newton was the first to explain why all objects fall down. He put together two concepts that at the first glance have nothing in common: centripetal force and free fall. Centripetal force is the force that makes an object move in a circular path. A good example is the friction force between the tires of a car and a road when making a turn. Without the friction force pulling the car to the center of the curve the car would move in a straight line. That is why an attempt to make a sharp turn with "bold" tires or on a slick road very often ends up in the nearest ditch. Another example is an object swinging on a string. It will swing around only when there is a tension force in the string. Remove the tension and let it go, the object will fly in a straight line. But what does all of this has to do with gravity? Let's think about other objects constantly moving around us such as the Moon rotating around the Earth or the Earth and all the planets moving around the Sun. Obviously each of them experiences a force that pulls them to the center of the orbit. Newton called this force gravity. He explained that an object near the surface of the Earth experiences the same kind of force as the celestial bodies and that an object in a "free fall" is not actually free of force. If this object is dropped it will go straight down toward the center of the Earth. If the object is initially moving horizontally, gravity will curve its path toward the center of the Earth and the object will eventually fall down. Following this logic Newton predicted that at certain speed it is possible to send a cannon ball to orbit the Earth. In the 20th century his prediction was proven right and today life is unimaginable without artificial satellites orbiting the Earth.

Figure 2.1. Gravity in the solar system.

Newton derived an equation that describes how gravity woks. In the 19th century Adams and Le Verrier independently used this equation and Newton's Laws of Motion to model the Solar System. Their calculations concluded that there had to be another planet in the Solar System besides the seven known at the time. Both Adams and Le Verrier were able to predict the mass and the orbit of the unknown planet and Johann Galle used Le Verrier's prediction to find the missing planet: Neptune. Today we know well how gravity works; we can model it and learn more about the Galaxy and the Universe but just as Newton 300 years ago, we are still not sure what gravity really is. Isn't it amazing that we use gravity to send robotic spacecrafts to explore the most distant parts of the Solar System yet we are not exactly sure about the nature of gravity? The understanding of gravity unveiled the Solar System and then the Universe to humans. The understanding of gravity changed our perception of space from an unknown scary dark place to an environment we are part of that we can study and explore. Who knows what's to come.

Figure 2.2 below illustrates a famous set of drawings that Newton drafted when thinking about the possibility to send an artificial satellite in orbit around the Earth.

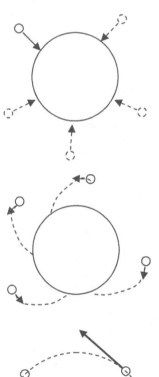

If one just drops an object elevated above the surface, the object will go straight down along a vertical line. The vertical line is the direction of gravitational pull – toward the center of the Earth.

If an object above moving above the surface of the earth has a horizontal speed, its trajectory will be progressively deflected by gravity in the direction of the gravitational pull – toward the center of the Earth.

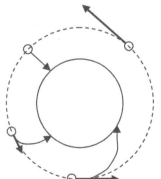

The trajectory of an object with greater horizontal speed will deflect less and therefore the object will land further. At some point the speed will be great enough for an object never to reach the surface and be in orbit around the earth.

Figure 2.2. Newton's concept of launching an object into orbit.

## *Activity 2.2:* "A Force of Attraction"

### "Ups" and "Downs" around the Earth.

Below you will find a graphic of the earth. Imagine that you are traveling all around the earth. Draw a schematic of a person standing on the earth first at the North Pole, then at the South Pole as well as on the west side and the east side of the globe.

**North Pole**

Schematic of
our traveler

**South Pole**

# "Rain, Rain, go to Spain!"

In this next figure you will find a cloud hanging over the North Pole. First draw clouds over the South Pole. Then it starts raining from both clouds; draw the water droplets coming down from the clouds.

## North Pole

## South Pole

## *Activity 2.3:* "How Do Objects Fall to the Ground?"

Do you think there will be a difference when a penny or a piece of paper the size of a penny is dropped? In both cases explain your point of view and formulate a hypothesis as to what will happen.

Your hypothesis: _____

_____

Conduct this experiment as a group of 4-5 students. You are given a stopwatch, a penny, identical pieces of paper (one per student), and a yard stick.

1. **Each student scrambles the piece of paper to different degree of tightness** (one student should leave the piece of paper "as is", flat).
2. One student will hold the yard stick to set the height from which the papers and pennies are dropped.
3. Another student will hold the timer and measure as accurately as possible the time that it takes for the penny and the papers to drop to the floor.
4. A third student will record the drop times.
5. Other students will do the actual drop of the penny and the papers.
6. Designate one student to coordinate the activity.

Drop successively the penny and the papers and record in the table below the time from the moment they are let go to the moment they hit the ground.

| Object | Description | Drop time | Comment |
|--------|-------------|-----------|---------|
| 1. | Penny | | |
| 2. | | | |
| 3. | | | |
| 4. | | | |
| 5. | | | |

Based on the results of your experiment revise (or not) your hypothesis and explain the difference if any between how a penny and piece of paper fall and why there might or might not be such difference.

Revised Hypothesis:_____

## How do object fall to the ground in a Vacuum?

In this activity, the instructor will repeat the penny and paper drop inside a vacuum tube.

What did you observe: _____

In your conclusion tell us what makes objects fall and what is the cause of the difference between how objects fall.

# *Activity 2.4:* "Modeling the Motion of a Rocket"

We are going to *model* the motion of a rocket by using a balloon. A model is not the real object (a rocket in this case) but some of the model properties are similar to those of the real object. A model helps us study more easily certain properties of the real object. This is especially useful when the real object is inaccessible.

The principle behind any rocket is propulsion motion. This could be explained in simple terms as motion due to recoil. When gas streams out of a rocket it produces the thrust that moves the rocket in the opposite direction. A rubber balloon can be used as an illustration or model of such process. Similarly, when air gets out from an untied balloon it produces thrust and the balloon moves in the opposite direction. The only difference is that the stream of air coming out of a balloon keeps changing directions and the balloon bounces from side to side. To direct the motion of the balloon we will use a light string drawn through a soda straw which is attached to the balloon by a piece of tape.

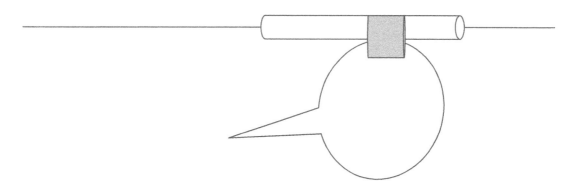

It is obvious that gravity will affect the motion of a balloon (rocket). But how large will this effect be? The answer to that question is the goal of this experiment. The answer must be quantitative. This means that during the discussion you will compare the distances balloons of same weight travel in the horizontal and the vertical directions. Then analyze the differences to see if there is a pattern. At the end, in conclusion, estimate the ratio of the change in distance for each additional weight due to gravity.

## Problem:

*How large will be the effect of gravity on a balloon's (rocket) travel?*

## Hypothesis:

_____

_____

_____

To answer this question a balloon needs to be released horizontally first to see how far the balloon will go while excluding the effect of the gravity. In the second experiment we will release the balloons vertically to maximize the effect of gravity (no support from the string in this case). The conclusion will be made based on the comparison of the distance reached in these two cases. For both experiments we will keep the same amount of the air in the balloon but we will increase the mass of the balloon by attaching increasing weights (to increase the force of gravity).

**Materials:** Straw, tape, spool of string, balloon, scissors, measure tape.

## Procedure:

1. Draw the string through the straw.
2. Blow 3 blows into the balloon but do not tie it, just keep it closed.
3. Attach the straw to the balloon by taping it over, closed end facing the spool.
4. The partner holding the balloon takes the spool and stretches the string 30ft.
5. While holding the string tight release the balloon.
6. Measure and record the distance traveled by the balloon.
7. Repeat steps 2 – 6 two more times and calculate the average.
8. Repeat steps 2 – 7 but each time tape increasing weights to the bottom of the balloon.
9. Repeat steps 2 – 8 using exactly the same weights but stretching the string vertically.
   (One of the partners needs to step on a chair)

## Experimental Results:

A. Horizontally.

Number of blows *(const)*: 3

| Mass attached to balloon, grams. *(manipulated.)* | Distance traveled by balloon, meters. *(responding)* | | | |
| --- | --- | --- | --- | --- |
| | Trial #1 | Trial #2 | Trial #3 | Average |
| | | | | |
| | | | | |
| | | | | |
| | | | | |

B. Vertically.

Number of blows *(const)*: 3

| Mass attached to balloon, grams. *(manipulated.)* | Distance traveled by balloon, meters. *(responding)* | | | |
| --- | --- | --- | --- | --- |
| | Trial #1 | Trial #2 | Trial #3 | Average |
| | | | | |
| | | | | |
| | | | | |
| | | | | |

## Discussion and Conclusion:

Discuss how the distance traveled by the balloon was affected by additional weights for the motions in both directions. Then compare the changes in vertical (max. gravity) and horizontal (min. gravity) distances traveled by the balloon as the same additional weights were added. Finally make a conclusion about the effect of gravity based on this comparison. Review your hypothesis. Make suggestions/comments as to how a real rocket must be designed.

# 2.3 Simple Machines

Work can be done more easily by using machines. Machines with a few or no moving parts are called simple machines. They are used to reduce the amount of force needed or to change the direction of an applied force. For example it is easier to pull down than to pull up against gravity. It is also easier to roll than slide against friction.

Although machines make it possible to use less force, they do not reduce the amount of work. To gain on the force one need to give up something. With simple machines when the force is reduced, the distance required to move the object is increased.

The "Golden Rule" of mechanic says that if you have advantage in force you will have a disadvantage in distance and vise versa.

$$\text{Effort} \times \text{Distance of effort} = \text{Resistance} \times \text{Distance of resistance.}$$

**Effort** is the force you apply to do the work.
**Distance of effort** is the distance you apply this force.
**Resistance** is the weight which needs to be lifted up or any other resistance.
**Distance of resistance** is the distance this weight needs to be moved.

There are six simple machines which will be represented in this unit:

*Inclined plane* – an inclined plane is a ramp. Instead of lifting a weight straight up one moves the object along a ramp gradually rising to the necessary height. Ancient Egyptians used inclined planes to elevate enormous stone blocks while building the pyramids.

*Wedge* – a wedge is two inclined planes put together to form a sharp edge. It helps lift something up or split it apart.

*Wheel and axel* – it is easier to roll than to slide because only a very small portion is touching the ground which reduces friction. Instead of sliding an object over a surface one moves it on the axel attached to the rolling wheel.

*Lever* – is a beam free to move from side to side with a fixed support location somewhere along the stick. A lever rotates around this point called the fulcrum. Archimedes said: with the right amount of leverage one can even lift up the Earth.

*Pulley* – is a wheel with a groove on which a rope or cable can be located. A pulley can be fixed or movable. If one uses one wheel it will be a single pulley, if two pulleys are used we have a double pulley system and so on.

## Chart 2.3       "Types of Simple Machines"

| Name of machine | Diagram | Main advantage of the simple machine |
|---|---|---|
| Inclined Plane | Effort | The longer the inclined plane, the more gradual the slope, so the less force is required to move an object upward |
| Lever | Effort / Arm of Effort / Resistance | The longer the effort arm, the less force is needed to lift the resistance weight |
| Pulley | Effort / Movable pulley / Effort / Fixed pulley | A fixed pulley only changes the direction of the force. You pull down instead of pulling up. A movable pulley reduces the effort by half because two sections of the rope support the pulley. But also the rope will need to be pulled twice the distance |
| Wedge | | A wedge makes it easier to split an object or raise a side of the object |
| Wheel and Axle | Effort | A small force at the wheel moves a larger weight (resistance). The larger wheel compared to the axle, the greater gain in force |

# *Activity 2.5:* "Lift it up!"

In this activity you will lift blocks of various masses with and without the help of simple machines. As you can imagine, simple machines will make the work easier but can you compute the difference and what are you giving up?

**A. The Inclined Plane.** Take a spring scale and make sure it is measuring weights accurately. Use the spring scale to lift a block straight up. Then elevate the block to the same height but this time by pulling it up along an inclined plane.

Are the readings of the scale the same? _____

Lift the same block by pulling it up along the ramp of different steepness. Record the scale readings in the table below. Keep the ramp distance the same while changing its steepness.

Block is lifted straight up

Block is lifted along the inclined plane

| Scale reading when block is lifted straight up. (resistance) | Height of slope, cm. (dist. of resist) | Scale reading when block is moved along the incline plane. (effort) | Length of slope, cm. (dist. of effort) |
|---|---|---|---|
| | | | |
| | | | |
| | | | |
| | | | |

What can you state about the amount of effort (force) required to move an object up a ramp as the ramp becomes steeper?

_____

_____

What can you state about the amount of work (force x distance) required to move an object up a ramp as compared to straight up?

_____

_____

Can you apply the "Golden Rule" to this case?  Compute and compare.

Effort × Distance of Effort  =  Resistance × Distance of Resistance.

---

**B. The Lever**: First use the spring scale to lift a block straight up.  Then elevate the same block to the same height by using a lever and pulling downward with the spring scale.

Are the readings of the spring scale the same?     _____

Lift the same block several times by applying the effort at different distances from the fulcrum. Start with a distance of half the distance from the block to the fulcrum then gradually increase it. Record the lengths of the arms and the readings of the scale for each case in the table below.

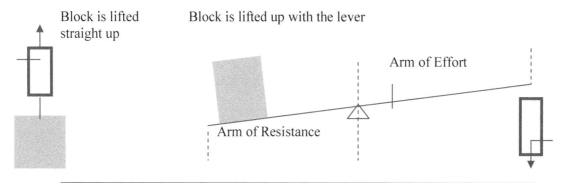

| Arm of effort, cm | Weight of block, g. (Resistance) | Arm of resistance, cm. | Effort required to lift resistance, g. |
|---|---|---|---|
|  |  |  |  |
|  |  |  |  |
|  |  |  |  |
|  |  |  |  |
|  |  |  |  |

What can you state about the amount of effort (force) required to move an object by a lever as the arm of effort becomes longer?

_____

How does the data from the experiment correlate with the "Golden Rule"

Effort × Distance of Effort  =  Resistance × Distance of Resistance.

_____

_____

A seesaw acts as a lever. When two people of the same weight are on a seesaw, it will be balanced if they are both the same distance from the fulcrum. If one person is twice as heavy, where must the person sit in order for the seesaw to be balanced?

_____

_____

_____


**C. Pulley**. Use the spring scale to lift a block straight up.  Then elevate the block to the same height by using the pulley.

Are the readings of the scale the same?   _____

Block is lifted straight up

Block is lifted up through the pulley

Construct a movable pulley and repeat the experiment above.  Compare the results. How are fixed pulleys and movable pulleys different?

_____

_____

_____

_____

_____

# Topic 3

# Matter, Chemistry, Water & Heat

# 3.1 It's All About Matter...

Pick up a piece of paper or clay. Imagine breaking it in half. Then each half split in half again. And again until the remaining piece is so small that you can no longer split it with your fingers. Imagine then cutting it under microscopes of increasing magnification. How far can one go? Is there a limit to splitting materials in smaller and smaller parts? What is matter made of? Greek philosophers debated this question. The famous Philosopher Aristotle thought that all materials were composed of a combination of four basic elements: Earth, Water, Air and Fire. Small changes in the respective proportions of these four basic elements lead to an infinite variety of materials. In this model materials could be transformed into other materials, including the transformation of lead into gold which captivated alchemists for centuries. Another Greek Philosopher by the name of Democritus made the hypothesis that matter is composed of very small building blocks, invisible to the naked eye suspended in void. He called these minute building blocks "atomos" from the greek atomon or indivisible. It followed that all forms of matter were the result of various combinations of such atoms of different types. In Democritus model there is no possibility of turning lead into gold as the materials are made of different atoms or building blocks. Both theories of matter are more involved and had considerable influences for more than two thousand years.

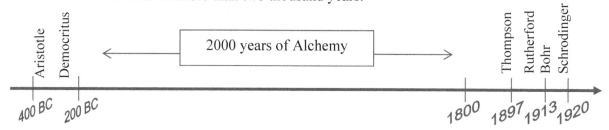

Figure 3.1. Development of human knowledge of the atom.

We had to wait until the end of the 1800s for scientists to finally start discovering experimentally the structure of matter. In 1897 J.J. Thompson was the first to measure the presence and characteristics of the electron, a tiny negative particle much lighter than a small atom. If the electrons were negative there was the need for something positive in the atom. Thompson and other scientists initially thought that the positive charges were spread throughout the atom. In 1911 Rutherford and colleagues observed that when a beam of alpha particles (helium nuclei) is shot through thin sheets of various materials, most of the alpha particles make it through with little or no deflection. In other words the atom has to be mostly empty space! In rare cases the alpha particle is reflected backward. The atom had to be made of a very small nucleus surrounded by mostly vacuum. But let's not forget about the electrons.

In Science one answer often leads to new questions and in this story the next question is: where exactly are these electrons around the nucleus? Bohr in 1913 came up with a planetary like model of the atom with electrons orbiting like planets at increasing distances from the atom. One of the key features of the Bohr model is that there is only a small set of positions/orbits that are available to the electrons, i.e. there is only a small precise set of energy levels that electrons can occupy. The model had to be further refined in the mid 1920s using the principles of quantum mechanics (Wolfgang Pauli) and including the wave description of the electron (Erwin Schrödinger). In this section we will familiarize ourselves with the structure of the atom, i.e. a very small but heavy nucleus made of protons and neutrons surrounded by a vast vacuum in which electrons move around in precise orbits around the nucleus.

# Chart 3.1:  "It's All About Matter…"

- The building blocks of **matter** are **atoms** and **molecules.**
- Atoms or molecules that compose matter are *in constant motion*.
- There is *empty space between atoms and molecules* composing matter.
- *Atoms and molecules interact* with forces of attraction or repulsion.

Molecules are made from atoms.
Atoms are composed of subatomic particles: **protons, neutrons and electrons.**
Protons and neutrons together form the *atomic center or nucleus*.
The nucleus is surrounded by electrons orbiting it at a relatively great distance.
Atoms differ from each other only by the *number of subatomic particles*.
The atom is the *smallest unit* that keeps *chemical properties* of matter.

Matter is usually in three different main states: **as solid, as liquid and as gas**.
In every state matter is composed of *the same atoms or molecules*.
The state of matter is determined only *by the amount of kinetic and potential energy* of the atoms or molecules in the matter. In different states matter has *different arrangements* of atoms or molecules moving at *different speeds*.

Matter in the state of **solid** has **the least** amount of energy.
Molecules are *barely moving* being *very close* to each other in well organized *order*. As a result solids keep both form and volume.

Matter in the state of **liquid** has **a medium** amount of energy.
Molecules are *moving fast* still being *close* to each other but not as orderly.
As a result liquids keep volume but not form.

Matter in the state of **gas** has **the greatest** amount of energy.
Molecules are moving *very fast* and are *very far* from each other in *no order*.
As a result gas keeps neither form nor volume.

Figure 3.2. Water in different states of matter. © MaxFX, 2009. Under license from Shutterstock, Inc.

**Exercise:** In the picture above you will find water in all three states of matter. Answer the questions below based on the picture.

1. Find on the picture liquid, solid and gaseous water.

_____

_____

2. Where on Earth do you think the picture was taken? Near the Equator, the North Pole or somewhere in between? Justify your answer.

_____

_____

_____

3. How will the picture change if temperature:

A. rises? _____

B. drops?_____

# 3.2 Building Blocks of Matter

Atoms are the building blocks of matter in all its phases. Atoms are made of protons, neutrons and electrons. Protons, neutrons and electrons have different properties. One of the properties that distinguish them is *electric charge*. The electric charge of a proton is positive, that of an electron is negative, and neutrons have no charge at all. Normally the number of electrons in an atom equals the number of protons so that the atom is neutral. Sometimes one or more electrons can leave their atom and join another one. An atom that has **lost one or more electrons** becomes a **positive ion** because now it has a positive electrical charge (it has more protons than electrons). An atom that has **gained one or more electrons** becomes a **negative ion** because now it has a negative electrical charge (it has more electrons than protons).

**Exercise:** a) What happens when the following pairs of particles are approaching each other? Do the particles repel or attract each other? Or neither? Use arrows to illustrate your opinion.

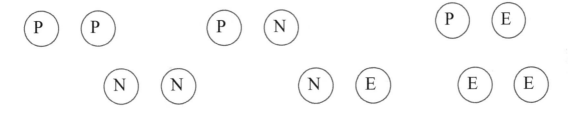

b) If you had to combine only two of these three particles to make an atom which combination would work and why would the other ones not work?

_____

_____

_____

The protons and neutrons of an atom form the nucleus of the atom. The nucleus represents most of the mass of the atom. The mass of an electron is insignificant in comparison with the mass of the proton or neutron. Based on this difference, the atomic mass (AM) roughly equals the sum of the number of protons (P) and the number of neutrons (N). Or simply AM = P+N.

Despite this rule different atoms of the same element can have different atomic mass because atoms often have several *isotopes*. An isotope of an atom has always the same number of protons and electrons but different number of neutrons. Therefore the atomic mass of the isotopes is different but their chemical identity and chemical properties are the same.

Table 3.1: "Building blocks of atoms"

| | Charge | Mass | Location | Role in the Atom |
|---|---|---|---|---|
| **Electron** | − | $5.5 \times 10^{-4}$ u | Orbiting at great distance from atomic nucleus | Chemical properties |
| **Proton** | + | 1.0 u | In atomic nucleus | Identity of element |
| **Neutron** | No Charge | 1.0 u | In atomic nucleus | Isotopes |

# Activity 3.1: "Decode the Periodic Table"

There are many types of atoms; in 1869 Dmitry Mendeleyev arranged atoms in rows according to the magnitude of their atomic weight. This table is called the periodic table of elements. Cells of the periodic table contain the main information about each element. You can find the name and atomic symbol of an element as well as its atomic mass and the number of each kind of particles this element is built from.

*Atomic Number* = number of protons = number of electrons

*Atomic Mass* = average mass including all the existing isotopes of the element

*Atomic Mass – Atomic Number* = average number of neutrons in all isotopes (round off AM)

1. Become familiar with the symbols of the periodic table of elements.

   Name of an element: _____

     Atomic number: _____

     Atomic symbol: _____

     Atomic mass: _____

   | Phosphorus |
   | :---: |
   | 15 |
   | **P** |
   | 30.97 |

   Choose another four elements in sequence from any part of the periodic table and gather the information about the name, atomic symbol, atomic number and mass of each element. Use this information to fill in the table below.

2. How many elementary particles of each kind do the selected elements have?

   | Name of element | # of protons | # of electrons | # of neutrons |
   | --- | --- | --- | --- |
   |  |  |  |  |
   |  |  |  |  |
   |  |  |  |  |
   |  |  |  |  |
   |  |  |  |  |

3. How are the atoms organized in the table? Find a pattern in placement of elements by Atomic Number and Atomic Mass.

4. Collect beads of different colors with the same sizes but different colors for protons and neutrons and a smaller size for electrons. Use the information from the periodic table to assemble on the tabletop atoms of Hydrogen, Helium, and Lithium. The protons and neutrons are bunched up in the center and form the nucleus and the electrons orbit around the nucleus (in reality far away from the nucleus). Compare the complexity of the atoms above:

_____

_____

_____

_____

_____

5. Using the periodic table, fill in the information missing in the table below:

| Name of Element | Chemical Symbol | Atomic Number | Number of Electrons | Atomic Mass | Number of Neutrons |
|---|---|---|---|---|---|
| Carbon | | | | | |
| | K | | | | |
| | | 11 | | | |
| | | | 17 | | |
| | | | | 79 | |
| | | 17 | | | |
| | | | | | 8 |
| | Au | | | | |

Figure 3.3. A pictorial history of the origin of the charge of elementary particles.

# 3.3 Where do all these Electrons Fit???!

As you are building bigger and bigger atoms, the neutrons and protons are simply assembled at the center of the atom but how will you place the growing number of electrons? There are very specific rules to place the electrons. The nucleus is surrounded by a series of **shells**. The electrons progressively fill up these shells. Shells are numbered in order of completion. Atom can not start to fill in the next shell until the previous one is complete. Within each shell electrons reside on different **orbitals**. Each orbital can host only up to two electrons (one spins clockwise and another spins counterclockwise). Each shell has different orbitals. The orbitals are labeled with letters **s, p, d, f, g, h**. Shell 1 has only one "s" orbital with a maximum of 2 electrons. Shell 2 can handle up to 8 electrons. Besides the "s" orbital with 2 electrons Shell 2 can have up to three "p" orbitals with each a maximum of 2 electrons. Atoms can have one, two, up to a maximum of three "p" orbitals because our space is three dimensional. Shell 3 is similar to shell 2 with one "s" orbital and up to three "p" orbitals. The organization of the other shells becomes a little more complicated. The number of the **period** (or row in the table) indicates how many shells an atom has. All but the last shell must be completed. Looking at the **group** (or column in the table) to which an element belongs, one can easily find out which orbitals form the last shell. The number of electrons occupying the last shell can also be found from the periodic table. It is simply the number of the group. Electrons that reside on the last shell are called **valence electrons**. Valence electrons play an important role in chemical bonding. Atoms form bonds by the means of valence electrons trying to reach the state when all shells including the last one are complete. Noble gases are the elements that occupy the last group in the Periodic Table and have all their shells including the last shell complete. They are chemically non-active, they do not participate in chemical reactions, do not form bonds, and can not be part of any molecule.

## Chart 3.2: "Electronic Configuration of Elements"

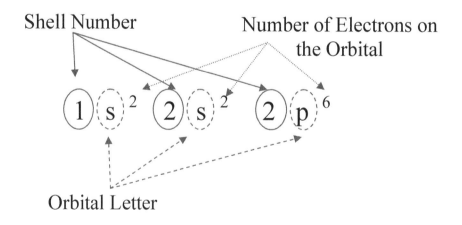

**Exercise:** Above you will find the electronic configuration of **Neon ($1s^2 2s^2 2p^6$)**. How would you know if it is a noble gas based on its e-configuration?

_____

# Activity 3.2: "Decode the electronic configuration"

1. Use Periodic Table to complete the table below:

| # of period | A<br>// # of elements in the period | B<br>// # of elements with only "s" orbital(s) in the last shell | C<br>// # of elements with both "s" & "p" orbitals in the last shell |
|---|---|---|---|
| 1 | | | |
| 2 | | | |
| 3 | | | |

Is there any difference in number from column B for all periods? How can you explain that?

_____

_____

_____

Is there any difference in number from column C for all periods? How can you explain that?

_____

_____

_____

Compare the number from column A with the sum of numbers from columns B, and C for all periods. Do you see a pattern? How can you explain that?

_____

_____

_____

_____

Period #4 has atoms that fit up to 10 electrons on "d" orbitals. Without looking at the Periodic Table predict how many electrons reside on the last shell of Kr, the noble gas of the fourth period. Use the Periodic Table to check your prediction. Give a simple rule how to find out the number of electrons on the last shell of an element and how to find out the number of electrons an element needs to complete its last shell.

_____

_____

_____

2. Find in Periodic Table the elements Carbon and Sodium. Write down the atomic configuration of these two atoms in terms of shells (1,2,..) and orbitals (s,p,…). Which one has the larges number of valence electrons?

Carbon:

Sodium:

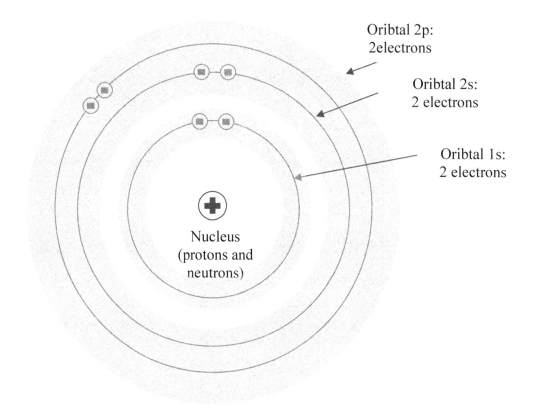

Oribtal 2p:
2electrons

Oribtal 2s:
2 electrons

Oribtal 1s:
2 electrons

Nucleus
(protons and
neutrons)

3. What element is represented in the chart 3.3? Justify your answer.

_____
_____
_____
_____

4. On the periodic table locate the element Aluminum and record the following:

- Atomic Number: _____
- Number of protons:_____
- Atomic mass:_____
- Number of neutrons:_____
- Number of electrons:_____
- E - Configuration:_____

# Activity 3.3: "Decode the Molecules"

Use colored beads to compose on your table two atoms of hydrogen with the nucleus at the center and the electron around the nucleus. Assume then that the two atoms are approaching each other. As you observe the atoms getting closer to each other what is the first thing they see of the each other. Based on this observation what component(s) of the atom is (are) important to determine what happens when the two atoms "get in touch"?

_____

Two or more atoms can combine to form a larger type of building block for matter: "a molecule". Not all atoms form molecules. Electrons residing on the last shell of the atom are responsible for atoms bonding into molecules. The atoms in column 18 are known as the noble gases. They are very stable because all their shells are full. Therefore they do not bond to form molecules with any atom. There is no a single molecule that include any atom from 18 period.

1. Place the following elements in the appropriate box below: Al, Ar, Au, Cu, $Cl_2$, Fe, $H_2$, He, Kr, Ne, $N_2$, $O_3$, $O_2$, Zn.

| Do not bond | Bond but do not form a molecule | Bond into a molecule |
|---|---|---|
| _____ | _____ | _____ |
| _____ | _____ | _____ |
| _____ | _____ | _____ |
| _____ | _____ | _____ |
| _____ | _____ | _____ |

2. Use Periodic Table to find information to complete the chart below.

| | F | O | Na | Mg | Ne |
|---|---|---|---|---|---|
| Number of electrons on the atom's last shell. | | | | | |
| Does the atom need to **remove** or **add** electrons to complete its last shell? | | | | | |
| How many electrons do the atoms need to transfer to complete their last shell? | | | | | |
| What kind of ion ("+" or "-") do the atoms become as the result of the transfer? | | | | | |
| What is the charge of this ion as a result of the transfer? | | | | | |
| How many electrons remain on the last shell of the atom after the transfer? | | | | | |

73

Among the five atoms in the table above which ones will bond into a molecule and which ones will never bond into a molecule. Explain **Why**.

Will form a molecule:                                             Explanation:

_____&_____                    _____
_____&_____                    _____
_____&_____                    _____
_____&_____                    _____

Will never form a molecule

_____&_____                    _____
_____&_____                    _____
_____&_____                    _____
_____&_____                    _____

3. From which respective parts of the periodic table are the components of a molecule more likely to come from?


4. Let's now look at other examples of molecules. In the table below you will find the name of several molecules Water, Carbon Dioxide, Methane, Sulfuric acid, Table Salt, Hydrochloric Acid, and Ammonia. Search the internet for the chemical formula for each molecule and use the found information to complete the chart.

| Molecule | Formula | Number of Atoms<br>H  C  O  N  Na  Cl  S | Total # of atoms in the molecule |
|----------|---------|------------------------------------------|----------------------------------|
| Water | | | |
| Carbon Dioxide | | | |
| Methane | | | |
| Sulfuric Acid | | | |
| Table Salt | | | |
| Hydrochloric Acid | | | |
| Ammonia | | | |

# 3.4 Chemical Changes vs. Physical Changes

## Chart 3.3:   "Chemical and Physical Changes"

| PHYSICAL CHANGE: Change in shape | CHEMICAL CHANGE: |
|---|---|
| *No change in chemical composition*     Change in form <br>     Change in temperature <br>     Change in state of matter | *Any change which leads to the change in chemical composition.* |

**A physical change is reversible, a chemical change is not.**
**A chemical change results in the formation of one or more new substances.**

A piece of magnesium is chopped into small cut strips.
Strips are then heated up until they melt.
As a result we have a liquid instead of a solid but we still have magnesium.
- There is no change in chemical composition: a physical change

A piece of magnesium is added to a copper sulfate solution.
As a result we see red flakes but no longer the silver colored pieces of magnesium.
This indicates that magnesium has been substituted copper in the compound.
We no longer have copper sulfate but magnesium sulfate: a chemical change.

**Exercise:** Use highlighters of two different colors to color the circles inside the square. Then connect each box with the correct circle using an arrow of corresponding color.

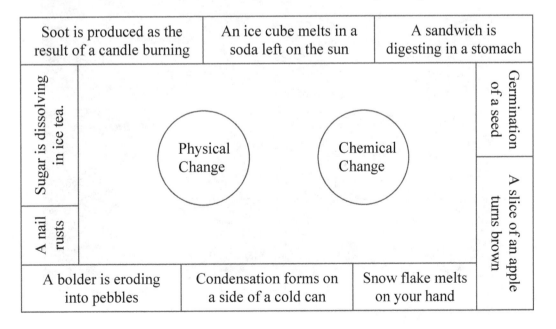

# *Activity 3.4:* "Growing Crystals"

To grow crystals you need a saturated solution of any salt. It may be a table salt (NaCl) or Epsom salt (MgSO$_4$), or even Borax (Na$_2$B$_4$O$_7$) found in the detergents section. To make a saturated solution dissolve as much salt as water will allow. When you can nolonger dissolve any more salt it means that saturation has been reached. The temperature of water for the saturated solution is a significant factor for the success of crystal growth. The hotter the better but always remember safety and be reasonable. To boil water is not recommended. Another not less important factor is the purity of the water and the salt. Distilled water is best for crystal growth. Remember to clean the container where you are going to make a solution and **do not use iodized table salt**.

When the saturated solution is ready you need to plant a seed for the crystals to grow. A variety of things can play the role of crystal growing centers. The most popular:
1. Charcoal - just place it in a shallow container filled with a saturated solution.
2. Cotton thread or wool yarn – tie it to a weight and drop it in the saturated solution.
3. Sponge (not rubber) or cardboard piece.

Now you need to make your own solution oversaturated to start the crystallization. Intense evaporation is one way but this may not be the best for some areas because of the high humidity. Reducing the temperature of the solution is the other. You may just leave the container to cool off to the room temperature on the countertop and this sometimes enough to start the process in a day or two. If not, try to place the container in the refrigerator or wait a little longer. Patience, do not disturb the process. Let it sit for at least two days and then look for the result. Good luck!

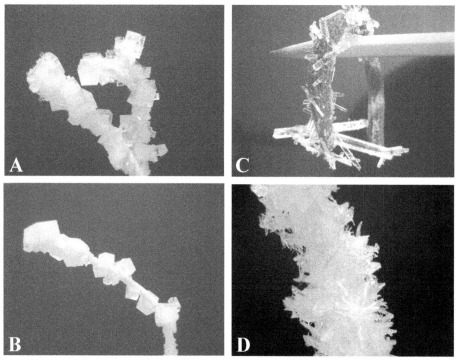

Figure 3.4. Photos of crystals grown by students using the above instructions (A & B, NaCl crystals, C & D, MgSO$_4$ cystals).

# 3.5 Pure Substances and Mixtures

Molecules can be composed of atoms of the same type or different types. If a molecule has only one type of atoms it is called an element. If atoms of the same kind bond without forming molecules as metals for example it is also called an element. Noble gases are elements as well. All elements are gathered in Periodic Table of Elements. A molecule formed from different kind of atoms is called a compound. But elements can be mixed up in the nature. And compounds can be mixed up too. As matter of fact elements can be mixed up with compounds as well. Then we are dealing with mixtures. Most of matter exists in the form of mixture. Mixtures are more common in the nature than either elements or compounds. The entire Earth is one large mixture.

## Chart 3.4: "Element? Compound? Mixture!"

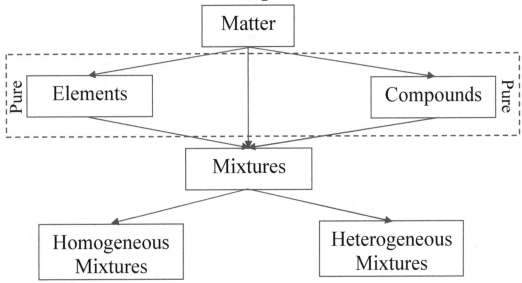

Properties of a compound are distinct from properties of each element forming the compound.

Components of a compound are present in very specific proportions (i.e. $H_2O$).

Components of a mixture retain their own properties and can be present in any proportion.

Components of a *mixture* can be *separated* by *a physical change*.

*A chemical change* needs to *separate* the components of a *compound*.

Atoms can not be separated by either physical or chemical means.

| ELEMENT | COMPOUND | MIXTURE |
|---|---|---|
| Is composed from *only one type* of atoms | Is composed from the *different types* of atoms | Is the *combination* of compounds and elements |
| ◆ Hydrogen $H_2$ | ◆ Water $H_2O$ | ◆ Air ($H_2$, $O_2$, $N_2$,$CO_2$) |
| ◆ Chlorine $Cl_2$ | ◆ Table salt NaCl | ◆ Soil |
| ◆ Copper Cu | ◆ Copper sulfate $CuSo_4$ | ◆ Milk |

## Activity 3.5: "Mixtures"

1. Mix in a transparent plastic cup salt, pepper and sand. Is it the result a mixture or a compound? If it is a mixture, what type of mixture is it, homogeneous or heterogeneous?

   _____

2. Write down a method to separate these items (if the result is a compound there will not be a simple method to separate back the components). Using water should be helpful.

   _____
   _____

3. If you add enough water to fill half the cup and stir the mixture what will happen to each substance?
   salt_____pepper_____sand_____

4. Now, how can you recover the 3 initial components?

   _____
   _____

E. What physical or chemical properties of the mixture components did you use to separate the mixture above?

   _____
   _____

## Activity 3.6: "Sort it all out"

For this activity you will need beads of two different colors and twist ties. If you tie together beads of the same color you will model a molecule of an elemental material. A few single beads will make a replica of an elemental material as well. But with beads of the different colors tied together you will get a compound. Mixing single beads of different colors without tiding them up or mixing them with compound "molecules" is the way to illustrate the mixture. The only thing left is to place them into separate plastic bags and label.
It is a very simple but at the same time a very demonstrative activity.
But what to do when all materials are assembled and sorted out? Here are a few ideas:

1. Have a box with pictures (clipped from variety of old magazines) of different materials. Ask students to match a picture with a proper bag. They may clamp pictures to the appropriate bead-bag or cluster the pictures around the bead-bag.

2. Another idea is to ask students to write down a list of examples corresponding to the type of material in each bag (let say five examples per bag) and place it into the bag for display.

3. You may also take your students to the playground and let them observe the environment searching for examples. Then determine what type of matter got the most examples (of course it will be mixtures) and discuss why.

# 3.6 Chemical Bonds

*Chemical bonds* hold together atoms of the same or different elements forming a molecule or a piece of metal.  Major types of chemical bonds are:

**COVALENT BOND**       **IONIC BOND**       **METALLIC BOND**

## Chart 3.5:          "Chemical Bonding"

| Chemical Bond | How bond is formed | Examples |
|---|---|---|
| **METALLIC** | <ul><li>Metal atoms easily lose last shell electrons</li><li>As the result they become positive ions</li><li>A cloud of "free" electrons flow in between the ions holding a piece of metal together</li><li>Individual metals can be combined into alloys</li></ul> | **Fe (iron), Cu (copper)** |
| **IONIC** | <ul><li>Some atoms easily give up electrons and become positive ions</li><li>Some atoms aggressively gain electrons and become negative ions</li><li>When two such atoms meet electrons are transferred from one atom to another</li><li>The positive and negative ions bond together through the electric force</li></ul> | **NaCl (table salt), NaF (sodium fluorite)** |
| **COVALENT** | <ul><li>When atoms have a similar ability to retain electrons they share electrons instead of transfer</li><li>Only two electrons can be shared per bond</li><li>An atom can make multiple covalent bonds with one or more other atoms</li></ul> | **NH$_3$ (ammonia), O$_2$ (oxygen)** |

The type of bond two atoms form depends on how each atom retains its own electrons. Atoms of different elements hold their electrons differently. Electronegativity of an element describes the ability of an element to retain electrons:

- the more electronegative an element the stronger its hold on its electrons
- the less electronegative an element the weaker its hold on its electrons

The difference in electronegativity will determine the kind of bond between atoms:
- 1.7 difference or greater will lead to ionic bond
- 0.5 to 1.7 difference will lead to polar covalent bond
- less than 0.5 difference  will lead to covalent bond

# Chart 3.7: "Examples of Chemicals Formed through Ionic, Covalent and Metallic Bonding"

| Ionic Bonds | | Covalent Bonds | | Metallic Bonds | |
|---|---|---|---|---|---|
| Table Salt | NaCl | Oxygen | $O_2$ | Copper | Cu |
| HydrochloricAcid | HCl | Hydrogen | $H_2$ | Aluminum | Al |
| Bleach (Sodium hypochlorite) | NaOCl | Ammonia | $NH_3$ | Iron | Fe |
| Copper Sulfate | $CuSO_4$ | Carbon Dioxide | $CO_2$ | Magnesium | Mg |
| Baking Soda | $NaHCO_3$ | Water | $H_2O$ | Gold | Au |
| Lye (caustic soda) | NaOH | Isopropyl alcohol | $CH_3CHOHCH_3$ | Silver | Ag |
| Rochelle Salt (Potassium sodium tartrate) | $KNaC_4H_4O_6 \cdot 4H_2O$ | Common Sugar | $C_6H_{12}O_6$ | Mercury | Hg |
| Aluminum Oxide | $Al_2O_3$ | Vinegar (5% acetic acid) | $CH_3COOH$ | Steel (316 Stainless Steel) | Fe + 0.08% C + 2.0% Mn + 0.75% Si + 17.0 % Cr + 12.0% Ni + 2.5% Mo |

**Exercise:**   Use the Electronegativity chart below to predict what kind of bond the following atoms will form. Then try to write down the chemical formula for the compounds below:

| H | Li | Be | B | C | N | O | F | Na | Mg | Al | Si | P | S | Cl | K | Ca | Cr | Fe | Br |
|---|---|---|---|---|---|---|---|---|---|---|---|---|---|---|---|---|---|---|---|
| 2.1 | 1.0 | 1.5 | 2.0 | 2.5 | 3.0 | 3.5 | 4.0 | 0.9 | 1.2 | 1.5 | 1.8 | 2.1 | 2.5 | 3.0 | 0.8 | 1.0 | 1.6 | 1.8 | 3.0 |

| Atom 1 | Atom 2 | Electroneg. for Atom 1 | Electroneg. for Atom 2 | Difference in Electroneg. | Type of bond |
|---|---|---|---|---|---|
| K | Cl | | | | |
| Mg | O | | | | |
| C | Br | | | | |
| N | O | | | | |
| S | O | | | | |

# 3.7 Chemical Terminology

The chemical formula of a compound describes which elements make up the compound. It also tells us about the proportion of the participating elements. A charged group acting as a single unit in an ionic compound is called **Polyatomic Ion.** To indicate a polyatomic ion one needs to place it in parentheses. You will find below a list of common polyatomic ions.

| | |
|---|---|
| Ammonium | $(NH_4)^+$ |
| Carbonate | $(CO_3)^{2-}$ |
| Chlorate | $(ClO_3)^-$ |
| Cyanide | $(CN)^-$ |
| Hydrogen carbonate (or bicarbonate) | $(HCO_3)^-$ |
| Hydroxide | $(OH)^-$ |
| Nitrate | $(NO_3)^-$ |
| Nitrite | $(NO_2)^-$ |
| Permanganate | $(MnO_4)^-$ |
| Phosphate | $(PO_4)^{3-}$ |
| Phosphite | $(PO_3)^{3-}$ |
| Sulfate | $(SO_4)^{2-}$ |
| Sulfite | $(SO_3)^{2-}$ |

When naming an ionic molecule one needs to describe what ions are in this molecule. One starts with the name of the positive ion and finish with the name of the negative ion. Generally, if the negative ion is monatomic its name will end with the suffix – **ide**. For example NaCl has a positive sodium ion and a negative chlorine ion, so its name is sodium chlor*ide*. AlF$_3$ has a positive aluminum ion and three negative fluorine ions – its name is aluminum fluor*ide*. For ionic compounds there is no need to specify that there are three fluorine ions as the proportions are defined by the ionic charges of the participants and only one combination is possible. It is quite different for covalent compounds. The same elements can form a variety of covalent compounds with each element in different proportions. When naming a covalent compound one needs not only to have in the name the participating elements but also to identify the number of atoms for each element. For example, CO and $CO_2$ are composed from the same elements but in different proportions. A system of Greek prefixes helps to indicate the difference between the two above molecules. CO is carbon *mon*-oxide and $CO_2$ is carbon *di*-oxide. *Tri*- and *tetra*- are used to indicate three and four atoms respectively.

**Exercise:** Name the following compounds:

a) LiF; CaCl; Na(HCO$_3$); Cu(OH)$_2$; Al$_2$O$_3$      b) NO; NO$_2$; N$_2$O$_4$; N$_2$O; SO; S$_2$O; SO$_2$

# 3.8 Acidic and Basic Solutions

There are different classifications of chemical compounds. A widely used classification is that of acids, bases and salts and is described below.

## Chart 3.7: "Acids and Bases" Chemical Compounds

| Base | Acid | Salt |
|---|---|---|
| • Acids release a hydronium ion when dissolved in water | • Bases release a hydroxide ion when dissolved in water | • A salt is an ionic compound. |
| • Acids have a sour taste | • Bases have a bitter taste | • Salt is the product of a reaction between an acid and a base or an acid and a metal. |
| • Acids neutralize bases. | • Bases neutralize acids. | |
| • Acids corrode active metals | • Bases denature protein. | |
| • Acids change the color of some substances. | • Bases change the color of some substances | • As minerals salts are essential in the diet. |

**Chemical Properties:** An acid and a base react to make a *salt* and water. A salt is an ionic compound that could be made with the negative ion of an acid and the positive ion of a base. Water is an example with the hydrogen ion of the acid and the hydroxide ion of the base uniting to form $H_2O$. When an acid reacts with a metal, it produces a compound with the positive ion of the metal and the negative ion of the acid. Hydrogen gas is also produced in that reaction. Even the metal gold, the least active metal, can be dissolved by a mixture of acids called 'aqua regia,' or 'royal liquid'. Strong bases that dissolve well in water, such as sodium or potassium lye are very dangerous to living organisms because they react with proteins and a great amount of the structural material of living organisms is made of proteins. The "slippery" feeling on hands when exposed to a base comes from such caustic reaction. To avoid serious damage to flesh caused by strong bases and acids, make sure you follow safety procedures.

**Acid/Base identification:** Litmus is one of a large number of organic compounds that changes color when a solution changes acidity. Litmus turns red when in contact with acid and blue when in contact with base.

**Acids, bases and food:** The word 'sauer' in German means acid and is pronounced almost exactly the same way as 'sour' in English. (Sauerkraut is sour cabbage, cabbage preserved in its own fermented lactic acid) Stomach acid is hydrochloric acid. Acetic acid is the acid ingredient in vinegar. Citrus fruits such as lemons, oranges and limes have citric acid in the juice. There are very few food materials that are alkaline, but those that are taste bitter. It is even more important that care be taken in tasting bases. Tasting of bases is more dangerous than tasting acids due to the property of strong bases to denature protein.

TASTING LAB CHEMICALS IS NOT PERMITTED BY ANY SCHOOL.

## Activity 3.7: "Acidic or Basic? Can Color Tell?"

In this activity we will explore the concept of acidic and basic solutions and discover which category common solutions belong to.

Get an eye dropper, a cup of water to clean an eye dropper, an empty plastic cup for the waste, a series of small plastic cups for the chemicals, a white sample plate, blue and red litmus paper strips, pH strips, and a pair of safety goggles.

Fill the set of small cups half full with: Cabbage Juice, Vinegar, Baking Soda solution, Distilled Water, Salt Water, Lemon Juice, Aspirin solution, Tummy tablets solution, Alcohol, Windex.

Use masking tape to label each liquid.

**Experiment 1:** Place three droplets of liquids from the table below in each of 6 holes on the sample plate. Make a small mark so you can identify the chemicals. Then add three droplets of cabbage juice to each liquid. Notice the change in color for each liquid. Classify the chemicals based on the final colors of the mixtures. Feel free to invent names for each category.

| Chemicals | Original Color | Color after mixing in cabbage juice | Category |
|---|---|---|---|
| 1. Vinegar | | | |
| 2. Baking Soda solution | | | |
| 3. Water | | | |
| 4. Lemon Juice | | | |
| 5. Windex | | | |
| 6. Alcohol | | | |

**Experiment 2:** Which chemical above would work the best to turn all your mixed solutions in the sample tray to the green color? _____
Test your answer. Avoid overfilling the sample tray by removing some liquid with the eye dropper if necessary.

**Experiment 3:** Which chemical would work the best to turn all your green mixed solutions in the sample tray to the pink color? _____
Test your answer. Avoid overfilling the sample tray by removing some liquid with the eye dropper if necessary.

*Experiment 4:* Based on your results in experiment 3 write in the table below the names of the test liquids group by group. Add the names of the other liquids in the small cups to the list below. Place these other liquids names with the group you think they will fit best. In each case discuss why you picked the group (you will have to do some educated guessing). Then test each substance in the cups with both pink and blue litmus paper by dipping the paper strips into each substance. Record your observations of change in color in the Litmus Paper.

| Name of test chemical | Change in Paper Color for the Pink Litmus Paper? | Change in Paper Color for the Blue Litmus Paper? | Category |
|---|---|---|---|
| 1. | | | |
| 2. | | | |
| 3. | | | |
| 4. | | | |
| 5. | | | |
| 6. | | | |
| 7. | | | |
| 8. | | | |
| 9. | | | |
| 10. | | | |
| 11. | | | |
| 12. | | | |

Discuss the difference between testing the solutions with cabbage juice and litmus paper:

_____

_____

_____

_____

*Experiment 5:* Based on the previous results write below the names of the test liquids in order: acidic, neutral, and basic. Dip a strip of pH paper in each of the liquids. Read the pH papers following the instructions on the package and assign a pH reading for each substance. Record the results in the table below specifying if the substance is neutral, acid, or base.

Compare the results of this test with your previous experiments.

| Substance in predicted order: acidic, neutral, basic | pH paper reading | Acid, Base or Neutral? | Does it match with previous testing? |
|---|---|---|---|
| 1. | | | |
| 2. | | | |
| 3. | | | |
| 4. | | | |
| 5. | | | |
| 6. | | | |
| 7. | | | |
| 8. | | | |
| 9. | | | |
| 10. | | | |
| 11. | | | |
| 12. | | | |

*Experiment 6:* In 2 small cups prepare the following solutions:

   (1) 10 drops of vinegar – no water

   (2) 2 drops of vinegar – 8 drops of water.

Which solution do you think is the most acidic and why?_____

Test both solutions with pH paper. Does the pH paper match your previous answer?

_____

*Experiment 7:* In 2 small cups prepare the following solutions:

   (1) 10 drops of baking soda solution – no water

   (2) 2 drops of baking soda solution – 8 drops of water.

Which solution do you think is the most basic and why?_____

Test both solutions with pH paper. Does the pH paper match your previous answer?

_____

*Experiment 8:* Put 5 drops of cabbage juice in 3 different cups. Put 10 drops of vinegar solution in those same compartments.

(a) count how many drops of Windex it takes to neutralize (change to purple) the vinegar in the first cup: ____

(b) count how many drops of baking soda solution it takes to neutralize the vinegar in the second cup:____

(c) count how many drops of Liquid Plumbing it takes to neutralize the vinegar in the third cup:____

Which basic solution is the most basic: _____

Which basic solution is the least basic: _____

*Experiment 9:* Design with your group a similar activity as (8) but using a set of acidic solution. Write down the procedure then swap your procedure with that of another group. Follow the procedure of the other group and discuss the educational value of their procedure. Write your main observations and two recommendations as how they can improve the procedure.

# 3.9 Water: An Essential Molecule

Water = $H_2O$. In other words water is made of molecules composed of 2 parts hydrogen and 1 part oxygen. It looks just like any other molecule formed by covalent bond but water is most important to our lives. Actually without water and its unusual properties life as we know it would likely not have developed. So what is so special about water and why is water so important to our lives?

## Chart 3.8:    Water: A Very Important Molecule

**Water: Around our planet**

- 71% of the Earth's surface is covered with water

- Most of the water (97%) is in the oceans and is salty

- Most of the fresh water is in solid form in the ice caps and glaciers

- Only 2% of earth's water is in rivers and lakes

- Water is the only element that exists naturally in all three states of matter on earth

**Water: Supporting Life**

- Water is said to be the base for life

- Many important chemical reactions in plants need water in order for chemical reactions to occur

- Over half of a human body is made of water.

- Animals and humans need water for the same reason and can only drink fresh water

- Water is a very important nutrient. You should drink 6-8 glasses of water daily

- Water is the only substance known to have higher density in liquid state then in solid. That is why ice floats which makes natural water reservoirs freeze from the top down keeping fish alive through the winter months.

## *Activity 3.8:* "Water: the 'Mickey Mouse' molecule?"

Try to understand how water molecules stick to each other by drawing two additional water molecules in the diagram below around the one already drawn.

***Hint***: the hydrogen atoms are bonded to the oxygen atom through covalent bonds. Because an oxygen atom has a greater nuclear charge than a hydrogen atom, the shared electrons in each bond are more attracted to oxygen than to the hydrogen atoms.  For this reason the oxygen side of each bond is slightly negative and the hydrogen side is slightly positive. Draw a positive sign '+' on the Hydrogen side and a negative sign '-' on the oxygen sign and use this information to draw the other molecules.

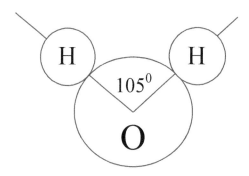

The water molecule has a positive and a negative side (is polarized) but what is the total charge of the molecule?_____

## *Activity 3.9:* **"Is Water "Sticky"?"**

In this activity we will explore the behavior and shape of water droplets to understand how water droplets form, their shape and how they interact.

Gather waxed paper or waxed sample glass; clean sample glass; medicine dropper; pencil

1. Use the dropper to put several water droplets of different size on the waxed paper or waxed glass, keeping each separate from the others. Examine the water drops.

- Do all of them have the same shape? What is it?

_____

_____

2. Carefully touch a drop of water with the pencil lead. Use magnified glass to observe how the drop of water responds.

- How did the surface of a drop of water changed when touched?

_____

- Do the water molecules seem to be attracted to the pencil lead or to each other?

_____

3. Push one of the drops around with your pencil and observe its behavior. Push two drops together, then three….

- Do the water molecules seem to be more attracted to the waxed paper or to each other? Is water sticky and if so to why?

_____

4. Use the dropper to put several water droplets on the clean sample glass, keeping each separate from the others. Examine the water drops.

- What is their shape?

_____

- Compare the shape of the droplets on the clean glass surface with the shape of droplets on the waxed surface? Is water sticky? and if so to what?

_____

_____

_____

Figure 3.5. Water droplets on a banana leaf.

## *Activity 3.10:* **"Fitting Water Droplets on a Penny"**

Let's further explore the consequences. How many drops of water will fit on a penny? Discuss this question with your group and write down a prediction for each group member. Then let each group member use an eye dropper to delicately place water droplets on a penny until the bubble bursts and the water spills outside of the penny. Count the droplets placed on the pennies.

| Group member's name | Prediction | Actual |
|---|---|---|
|  |  |  |
|  |  |  |
|  |  |  |
|  |  |  |
|  |  |  |

Explain what *property of the water molecule* allows you to place so many droplets on a penny:

_____

_____

_____


Based on experiment above predict and test how many paper clips (coins) can be placed into a full cup of water and not splash a drop? Test your prediction.

| Prediction | Actual |
|---|---|
|  |  |

We have learned through the past activities that water molecules have a negative charge on the oxygen atom side and a positive charge on the hydrogen's side. Keep in mind that the overall charge of the molecule is zero (neutral). Water molecules attract each other with the oxygen side of one molecule attracted to the hydrogen side of another.

When molecules are attracted to each other properties such as the water "stickiness" can result.

Here are a few additional definitions:

*Adhesion* – the attraction of unlike molecules to each other.

*Cohesion* - the attraction of like molecules to each other.

*Surface tension* – the attraction of the top layer of water molecules to each other.

**Exercise:** to test the concept of surface tension, take a paper clip, spread the parts of the paper clip and try to gently place it on the water surface. If you are successful, observe and describe the shape of the water surface below the paper clip.

_____

_____

_____

Once the paper clip is floating, add detergent to the water. Observe and describe what happens.

_____

_____

_____

**Exercise:** Surface tension makes the water surface flexible. The surface can act like a net and support objects of small weight such as the paper clip of the previous exercise. Insects use surface tension to run over the surface of water. Their legs do not break the surface. In the space below draw an insect walking on water with particular attention on the water surface.

# *Activity 3.11:* "How Substances Dissolve"

**A VERY SAD STORY......**

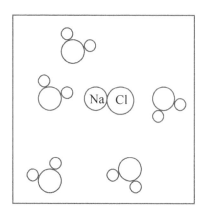 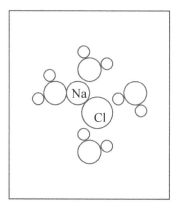 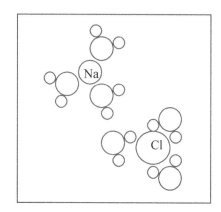

1. Once upon time a molecule of sodium chloride (NaCl) hold by an ionic bond fell into a glass of water.

2. Polar water molecules surrounded NaCl. Positive hydrogen sides moved towards the Cl ions. Negative oxygen sides moved towards the Na ion. Water molecules pull Na and Cl away from each other.

3. Finally, unable to resist, the sodium chloride molecule surrendered and was split apart...
A water solution of sodium chloride or simply salty water is the result of this battle.

Based on the story above, answer the following questions:

a. What is *the solution*? _____

b. Which one of the two components is a *solvent*? _____

c. Which one of two components is a *solute*? _____

d. When does a solution become saturated?

_____

e. Water is often called the universal solvent. Why?

_____

f. Make your own story about dissolving two crystals of sugar. Let one crystal dissolve in cold water and another one in hot water.

_____

_____

_____

## *Activity 3.12:* "Float or Sink? It is a Matter of Buoyancy!"

How to determine if an object will float or sink?

*Your Hypothesis:*

_____

_____

*Materials:* Plastic containers (or other containers that can be sealed), sand, cotton, pepper, steel paper clips, graduated cylinders or graduate beakers.

*Experiment 1:* Measure and record the volume of each container. Fill one of them with sand, another with cotton, then pepper, steel paper clips, and the last with any other available material. Twist the lid to make the containers waterproof. Predict which filled container will float and which will sink. Then weigh each container and compute the overall density of each container (mass of container divided by the total volume of the sealed container). Take a graduated beaker and fill it with about 150 ml of water to test your prediction.

The density of water is 1g/ml. Compare the density of *floaters* and *sinkers* to that of water.

| **Material** filling the container | **Prediction:** sink or float | **Mass** of filled container | Container **volume** (50 ml?) | Overall container **density** | Sink or float **observation** | **Comparison** with density of water (</>) |
|---|---|---|---|---|---|---|
| | | | | | | |
| | | | | | | |
| | | | | | | |
| | | | | | | |
| | | | | | | |

Based on the results of this experiment can you state a principle to predict if an object will float or sink in water?

_____

_____

_____

*Experiment 2: Let's Verify!* Based on the results of Experiment 1, predict the maximum overall weight of a container partially filled with pennies that will still be floating. Once you have a prediction test with an experiment. Plan for at least 3 trials, with (1) a weight of pennies such that the container is barely floating, (2) just sinking and (3) right at your prediction. At each step measure the weight of the container, the average density of the container, the volume of water displaced and the weight of water displaced.

| Trial # | **Volume** of container | **Mass** of container partially filled with pennies | Overall container **density** | Sink or float **prediction** | Sink or float **observation** |
|---------|-------------------------|-----------------------------------------------------|-------------------------------|------------------------------|-------------------------------|
|         |                         |                                                     |                               |                              |                               |
|         |                         |                                                     |                               |                              |                               |
|         |                         |                                                     |                               |                              |                               |
|         |                         |                                                     |                               |                              |                               |
|         |                         |                                                     |                               |                              |                               |

*Experiment 3: "Floating Potato"* Calculate the density of a small potato.
How buoyant is it according to your calculation? Will it float or sink? Test it.

_____

Get the potato out of water. Then add table salt to water to make a saturated solution and place potato back in. Is there any change in buoyancy of the potato? Describe it.

_____

What can explain this change?

_____

Based on the results of this experiment can you state a principle to predict if an object will float or sink in liquid other then water?

_____

_____

_____

# 3.10 Heat and Temperature

"A cup of hot coffe", "How cold is it outside?"– we use the terms "hot" and "cold" in everyday situations. General implication of these terms is how it feels to our skin. Indeed our skin is a heat sensor but what does this sensor measure? Temperature or heat?

Let's see. $70^0$F is room temperature and by its definition indicates general human comfort in air. However $70^0$F water feels quite cold. If our skin was measuring temperature then why is there such a different sensation for the same temperature? And what is temperature anyway?

Temperature measures how fast atoms and molecules are moving. Their speed correlates to the amount of thermal energy matter has <u>but not during phase changes</u>. Why? Think about why a steam coming from boiling water causes more sever burn then the same amount of boiling water? Although they both have the same temperature of $220^0$F, a steam has more thermal energy than boiling water. Beside kinetic energy due to motion of atoms and molecules Thermal Energy also includes potential energy of particles, and that makes this difference. Steam is gas and water molecules are further apart in steam than in boiling water which gives more potential energy to gas. Same concept can be applied to the question: Is it more effective to fill a cooler with icy water or melting ice? Of course ice! It is solid and though at the same temperature as icy water it has less Thermal Energy due to lack of potential energy because of molecular arrangement.

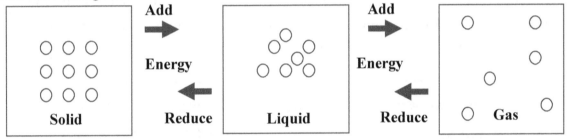

Figure 3.7. Thermal Energy is Kinetic and Potential Energy of all molecules in the object. The faster they move and the further they are apart, the more thermal energy the object has.

Heat is Thermal Energy in transit from one object to another. If it is not a phase change, it is correct to say that the temperature of an object is rising when heat is being transferred to it. It is also correct to say that when heat is transferred from an object its temperature drops (if it is not a phase change). But Thermal Energy can be changed not only by the heat transfer but by work or by energy conversion. A saw blade gets hot after cutting wood or car tires smoke after activating the emergency break. Inside a human body chemical energy stored in food gets transformed into thermal energy keeping our temperature above the temperature of surroundings at constant level.

Our judgment of "hot" or "cold" comes from our skin detecting the change in body temperature. When we touch a hot object our skin gains thermal energy and its temperature rises; when we touch a cold object our skin loses thermal energy and its temperature drops. Different materials conduct heat differently effecting how fast or slow temperature changes. This property of the material is called Thermal Conductivity. Styrofoam has low thermal conductivity and that is why it is better to make a coffee cup out of Styrofoam than aluminum which has a high thermal conductivity. By touching the inner and outer surface of the hot coffee cup you can notice a difference in temperature. Materials with low thermal conductivity make good heat insulators.

# Chart 3.8: "It's All About Thermal Energy..."

- Atoms or molecules which compose a substance are in *constant motion*. This means they have *Kinetic Energy*.
- Atoms or molecules which composed a substance *interact with each other*. This means they have *Potential Energy*.

**The total of Kinetic and Potential Energies of atoms and molecules composing the matter is** *Thermal Energy*

Temperature is the <u>indicator</u> not the measure of Thermal Energy of a substance.

At high temperature a substance has more Thermal Energy than at low.

A change in temperature of a substance indicates the change in Thermal Energy.

Thermal Energy changes without change in temperature during phase changes.

There are two ways to change the Thermal Energy of a substance.

**WORK**                    **HEAT** (energy in transit)
                            **TRANSFER**

Thermal energy of the substance *increases* when work is done *toward the substance*

Thermal energy of the substance *decreases* when work is done *by the substance itself*

**Convection**
Transfer of energy by streams of fluid or gas

**Conduction**
Transfer of energy from molecule to molecule

**Radiation**
Transfer of energy by electromagnetic waves

# Activity 3.13: "Ice Cream"

One of the most popular activities that address TEKS requirements on "Heat" topics is making ice-cream. It is relatively simple in organization and has a lot of science behind. It is very important to present its science in a correct way in order to avoid future misconceptions.

To make ice-cream you will need any kind of flavored milk (chocolate, raspberry, etc), ice cubes and salt. Here is a very simple recipe (per child):

> *1 small (quart) and 1 large (1galon) zip-lock bag*
> *1 plastic cup of milk (not full to prevent spills)*
> *3 plastic cups of ice*
> *2 plastic spoons of salt in a Dixie cup*

If you are doing this activity with 4th graders or older they should be able to just follow the instructions:

> *Place ice into large zip-lock bag; add salt*
> *Pour milk into small zip-lock bag and seal it*
> *Place the small bag with milk into the large bag with salt and seal it*
> *Shake the bags until you notice that milk become solid*

For the younger children you may want to distribute milk and ice in already sealed bags. Then ask them to add salt to the large bag, insert the small bag into the large and shake. You can accompany this with the music! Have fun!

The science behind this activity is very simple:

A) Milk transfers its thermal energy to the Ice because Milk has a higher temperature then Ice and heat always flows from higher temperature to lower temperature. To secure the understanding of this concept you may measure the temperatures of the Milk and the Ice with children at the beginning.

B) Milk loses energy as its molecules are slowing down, passing to the solid phase. This concept needs to be explained in advance; understanding that gain or loss in thermal energy always stands behind phase changes is very important.

C) The most confusing part is to understand the role of Salt. Salt drops the melting temperature of water. So what? It means that water can no longer be solid at 32 F; ice must melt quickly and drop the temperature below 32F! What does ice need to turn into liquid water? Energy!! What can provide it? Milk!!
Thermal energy transfers from Milk to Ice due to the temperature difference and water is now at a temperature below the solidification point of milk. As the Milk loses enough energy it becomes solid. And what is the name of solid milk? Ice – Cream!!!

D) If you want to expedite the process add an extra spoon of salt.

# Topic 4

# Electromagnetism & Waves

# 4.1 Electrical Charge and Charging

Electricity is all about charges. Objects are typically neutral, i.e. no net charge, but can become positive or negative. So how do you make a neutral object become charged? When you charge something you don't create positive or negative charge. To charge means to transfer electrons. Very often electrons are transferred from one object to another. What one object gains the other one loses. The gain of electrons leads to the negative charge. The loss of electrons leads to the positive charge. Although it may sound controversial not protons but electrons are responsible for the positive charge of an object because protons usually cannot be transferred and an object becomes positively charged if it loses electrons.

Different materials require different amount of work to take an electron away from an atom. The ability to retain electrons is referred to as the electronegativity of the material. More electronegative materials retain electrons better. For example, electrons are held more strongly in plastic than in your hair therefore plastic is more electronegative then hair. When you comb your hair, the plastic comb takes electrons from it. That makes a comb negatively charged because of an excess of electrons and hair positively charged because of a deficit in electrons.

Below materials are listed in order of increasing electronegativity. Materials easily giving up electrons listed first and materials holding on to their electrons more strongly last.

## Static Electricity / Charging Materials

- Human Hands (usually too moist though) (very positive)
- Rabbit Fur
- Glass
- Wool
- Lead
- Silk
- Aluminum
- Wood
- Hard Rubber
- Gold
- Polyester Styrene (Styrofoam) or Polypropylene Vinyl (PVC)
- Teflon (very negative)

**Exercise:**

A) What kind of charge does a glass wand acquire if you rub it with a piece of silk?

_____

What kind of charge does the piece of silk acquire? _____

Which one of the materials has larger charge if any (the difference between number of protons and electrons)? _____

B) What kind of charge does a PVC pipe acquire if you rub it with a piece of silk?

_____

What kind of charge does the piece of silk acquire? _____

# Chart 4.1:      "It's All About Charge..."

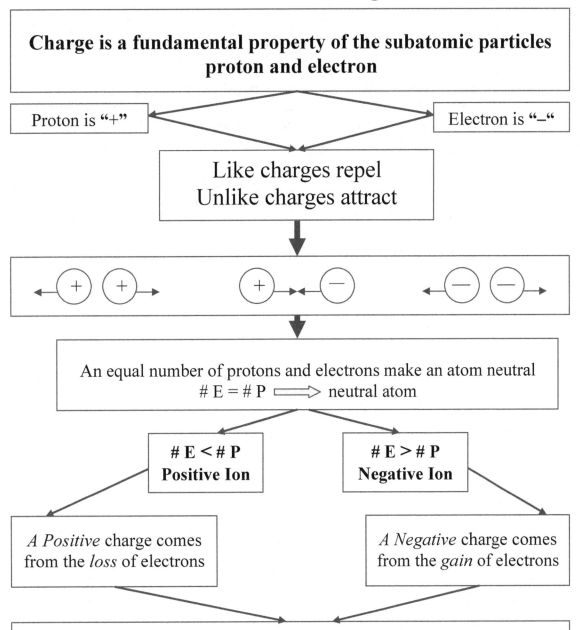

Charge is a fundamental property of the subatomic particles proton and electron

Proton is "+"                    Electron is "–"

Like charges repel
Unlike charges attract

An equal number of protons and electrons make an atom neutral
$\# E = \# P \Longrightarrow$ neutral atom

$\# E < \# P$
**Positive Ion**

$\# E > \# P$
**Negative Ion**

*A Positive* charge comes from the *loss* of electrons

*A Negative* charge comes from the *gain* of electrons

- Inner electrons are attracted more strongly to the nucleus
- Outer electrons can be more easily lost from an atom
- If an atom loses an electron it becomes a positive ion with (+e) charge
- If an atom gains an electron it becomes a negative ion with (-e) charge
- An object is negatively charged (-) if it has an excess of electrons
- An object is positively charged (+) if it has a deficit of electrons

## *Activity 4.1:* "Electrical Charge and Charging"

An atom has an equal number of protons and electrons which makes it neutral.
*If an object is a combination of neutral atoms it is neutral itself.*

Electrons on the inner shells of an atom are attracted more strongly to the nucleus than electrons from outer shells. Therefore outer electrons can be lost from the atom more easily. If an atom loses an electron it becomes a ***positive ion*** with the charge *equal to the charge of one electron (+e)*.

What will be the charge of an atom if the atom loses two electrons? _____

What if the atom loses ten electrons? _____

Imagine a situation when a few atoms of object **A** have lost a few electrons.

What could you say about the charge of object **A**, is it still neutral? _____

What kind of charge if any has object **A,** positive or negative? Why?
_____

If electrons have been transferred from object **A** to object B what is the charge of object **B**? Why? _____

Rub your hair with a balloon. Human hair does not retain electrons very well but hard rubber does and in the process of rubbing some electrons will be transferred from your hair to the balloon.

What kind of charge has the balloon, positive or negative? _____

What kind of charge has the hair, positive or negative? _____

Predict what happens if you bring the balloon close to hair? Why?
_____

Check your prediction.

Ask your group members to rub their hair with a balloon. Then carefully try to bring two balloons close to each other. What happens to the balloons? Why?
_____
_____

How does your hair behave after the rubbing with the balloon? Why?
_____
_____

## Exercise:

Mike rubs his hair with an inflated balloon and looks at himself in the mirror. He is horrified; his hair stands up and out!! But why?

To answer to that question he recalls what his science teacher said about charges:

(a) unlike charges attract

(b) like charges repel

(c) electrons can move from one material to another when in contact

Mike scratches his head with the balloon one more time and suddenly says "Bingo! That is why!"

 What would be your answer?

_____

_____

_____

_____

Figure 4.1. Styling innovation: ballooned hair!

# 4.2 Electrical Polarization

Charge an inflated balloon by rubbing it with your hair. Then place the balloon near a piece of paper. The piece of paper is attracted to the balloon! How is it possible?

The piece of paper was neutral it didn't have a charge. Does it have a charge now? No, you place the balloon near a piece of paper but you did not touch it! The piece of paper is still neutral but **polarized**. The charge on the balloon disturbs the charge arrangement in the paper. As molecules of water have positive and negative ends, the molecules of paper have them too. So, the positive part of each molecule is attracted toward the negatively charged balloon. What about the negative part? It is repelled!! And the molecules will rearrange themselves in such way that the positive side of almost each molecule will face the balloon. As the result of this rearrangement one side of the piece of paper becomes positive and another becomes negative but as a whole it is still neutral.

You can actually see the polarization of water if the charged balloon is placed near a weak water stream coming from a facet.

## Chart 4.2:    "Phenomena of Static Electricity"

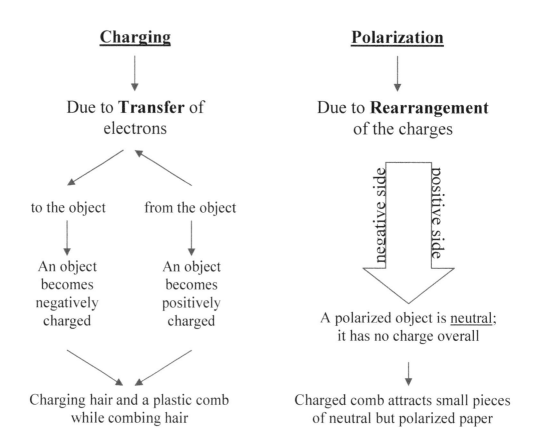

# 4.3 Electric Current and Electrical Circuits

Current Electricity is about moving charges while Static Electricity was about static charges (i.e. not moving). When charges collect on the surface of an object and remain there it is an example of static electricity. But when charges move through an object as a stream or current it is an example of electric current.

A material that allows such flow is called **conductor**. All metals are good conductors because metallic bonds have electrons that can freely move from atom to atom and between atoms. Such "free electrons" make it easy for charges to flow in metals.

A solution can also be a conductor. In solutions positive and negative ions can produce current by flowing in opposite directions. The human body is full of solutions which makes it a relatively good conductor. So, if you touch a charged object with your hand some charge will be transferred to you. But if you touch it with a dry wooden stick no charge will flow. Dry wood is **insulator** which means it doesn't conduct electrical charge.

Ionized gases (contain ions) can be conductors as well. A gas can be ionized if some force such as UV light removes electrons from the gas molecules. Such "freed electrons" and ions also can produce current by flowing in opposite direction.

A battery or dry cell has "+" charges concentrating on one end and "-" charges on the other end of the battery. If you connect the two ends of the battery with a metal wire (conductor) electrons will stream from the "-" to the "+"and we will have an electric current passing through the wire. If you disconnect the wire the current stops because air is an insulator.

# A Few Words about Electrical Safety

As with every science experiment safety comes first. For activities involving current electricity you should be particularly cautious in the choice of electrical energy source. The level of danger with electrical energy is related to the current passing through your body. You should therefore choose low electrical energy sources and if in doubt refer to a specialist. Activities involving static electricity such as rubbing a balloon are safe because although the electrical potential can be high they involve very limited number of charges. Activities involving low power batteries such as 1.5 V and 3V are safe as very little current would be generated in the human body. 9V batteries can be useful in experiments but more caution/supervision is needed. In general you should avoid activities using higher voltages and electrical power from wall outlets at least until High School when students are better able to judge the potential dangers.

## *Activity 4.2:* "Electrical Current"

**A: How to light a light bulb?** (Individual activity or by pairs at the most)

Use a small light bulb, a 1.5 V AAA battery and 1 wire only to find several ways to put together a circuit so that the light bulb is turned on.
Make schematics such of the different circuits that allowed you to turn on the light bulb.

What are the essential elements required for an electrical circuit?

_____
_____
_____
_____
_____

**B: Simple electrical circuit**

Once you have found several ways to turn on the light bulb, gather three other wires, a bulb holder and a switch to make your circuit more permanent and to be able to turn it on/off.
Put your circuit together and draw a detailed schematic of this circuit showing in particular how the light bulb filament inside the bulb is connected to the circuit.

## C: Electrical properties of materials

Modify your circuit such so that you have two wires with loose ends in the middle of the circuit. We will use this new circuit to characterize the conductive properties of materials. Objects of different known materials (material samples) can be found in a plastic bag. Close the circuit connection between the two "loose" wires with a material sample and check what happens to the light bulb. If the light bulb shines, electricity was able to pass through the material; if the light bulb does not shine, electricity was not able to go through the material. Before proceeding with the experiment, predict what materials will produce the least light (or no light at all) and which ones will produce the most and write down your predictions (no light at all, low light, bright light, etc). Record your results in the table below.

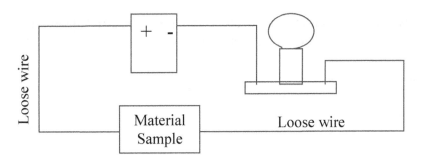

| Material no. | Material Tested | Prediction | Result (how bright is the bulb) | Comment on electrical conductivity of the material |
|---|---|---|---|---|
| 1 | Wood | | | |
| 2 | Aluminum | | | |
| 3 | Chalk | | | |
| 4 | Nylon | | | |
| 5 | Brass | | | |
| 6 | Steel | | | |
| 7 | Graphite | | | |
| 8 | Plastic | | | |
| 9 | Copper | | | |
| 10 | Glass | | | |
| 11 | | | | |

**D: Conductivity of water.**

Place the two loose wires in a cup filled with distilled water to see if the bulb will light up.

Is pure water a conductor or an insulator? _____

How about different water solutions?

Water – sugar solution: _____

Water – salt solution: _____

For the water-salt solution investigate either the effect of concentration, or the effect of distance between electrodes on electrical current.

Problem:

Hypothesis:

Experimental design:

Observation:

Conclusion:

# 4.4 Voltage, Amperage, and Resistance.

Table 4.1: "Voltage, Amperage, Resistance"

|  | Symbol | Characteristic of | Meaning | unit |
|---|---|---|---|---|
| **Voltage** | **V** | ***Battery*** | Work that will be done by the battery when moving a unit charge through a circuit | Volts, **V** |
| **Amperage** | **I** | ***Current*** | How much charge flows per second through the wire's cross section | Amps, **A** |
| **Resistance** | **R** | ***Wire*** | How much resistance in the wire to the passage of current | Ohms, **Ω** |

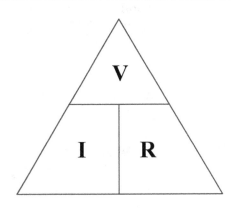

➢ Current <u>increases</u> with a larger Voltage or smaller Resistance

➢ Current <u>decreases</u> with a smaller Voltage or greater Resistance

How strong should be the current through a human body to do a significant damage? There is consensus that 0.1amp will definitely harm and possibly kill you. The question is what voltage will produce such a current? An interesting analysis of this can be found in online book www.allaboutcircuits.com, Volume I, Chapter 3, "Ohms Law (again!)".

Figure 4.2. Illustration that one should not take chances with apparently innocent and peaceful electrical power sources.

# *Activity 4.3:* "Parallel and Series Circuits"

Goal: To demonstrate the different ways of assembling an electrical circuit.

Materials: Two small light bulbs in the holders, a 1.5 V AAA battery and 6 wires.

There are two different ways to assemble an electrical circuit: in parallel and in series. Diagrams below illustrate both of them.

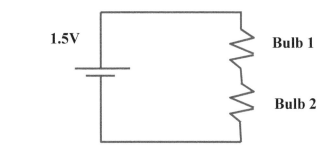

**A.** In *Series* bulbs follow each other along the same path. The bulbs are sharing the voltage of the battery.

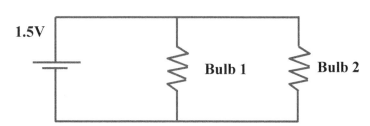

**B.** In *Parallel* the bulbs have independent paths to the battery connections. Each bulb is independently connected to the battery.

1) Assemble a series circuit following diagram A. Investigate what happens if you disconnect one of the bulbs.

_____

2) Assemble a parallel circuit following diagram B. Investigate what happens if you disconnect one of the bulbs.

_____

3) Compare the brightness of the identical bulbs in a series and a parallel circuit. In what type of connection do bulbs shine brighter?

_____

4) Based on your observation conclude in which (A or B) is there more current flowing through a single bulb. Why?

_____

5) What type of wiring is used in your home? Why?

_____

110

# *Activity 4.4:* "How to make a circuit maze"

1. Obtain two pieces of poster board the size you wish the maze board to be.

2. Using a hole-puncher, punch out several holes down the left side of the poster board. Repeat this down the right side of the poster board. It will now look like this:

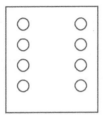

3. Lay this piece over the second piece of poster board and trace the holes. Take this second piece and cut strips of aluminum foil that will reach from the left column to the right column. Decide the direction of your first circuit.

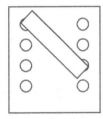

Perhaps you plan for your first hole in the left column to connect with the third hole of the right column, so make your foil strip the correct length. Place a wide strip of tape completely over the foil strip to insulate the metal.

4. Continue making your circuit pattern in this way. Be careful to cover each foil strip with a wide piece of tape such that they are all insulated from each other. Do not forget to keep track of the pattern so you will know what answer corresponds to what question.

5. When your circuits are complete, place the poster board with the punched holes on top of the circuits board. Now, only the foil dots will show.

6. Make labels for each pair of answers and questions. You can use mailing labels or just ordinary paper that then can be carefully glued or taped. You can use Velcro tape if you want to make sets of questions for different topics.

7. Place question #1 by the first dot in the left column and the answer to it by the third dot of the right column (for our example). Continue in the same way with the rest of questions and answers.

7. Tape or staple the two pieces of poster board together and your circuit maze is ready to use.

8. Now you need to make an indicator for the maze. Obtain two old markers, a Christmas Light bulb, two paper clips, 20cm of wire, and two AA or AAA batteries depending on the width of the markers.

9. Open each marker by pulling the cap out of the back of the marker then remove the leads and the ink stem. Place in a paper clip instead of each lead.

10. Insert a battery into each marker.

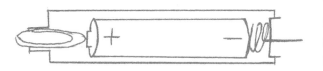

The battery in one marker needs to touch the paperclip by the "+" and the battery in another marker needs to touch the paperclip by the "-".

If you place batteries the wrong way they will be working against each other.

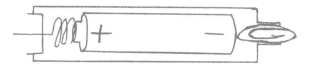

11. Make a hole in each marker's cap (a careful use of a hot nail makes it very easy) and insert the wire ends through these holes.

12. Twist the ends of the wire in small spirals so that they will not slide away then close each marker. Make sure there is a good connection between the battery and the wire spiral.

13. Strip plastic from the wire ends of a Christmas Light bulb. Cut the wire connecting the two markers in the middle and insert the Christmas Light bulb by twisting the wire ends together for better connection. Insulate the connection with any tape and you have your tester.

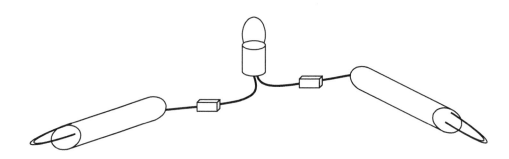

# 4.5 Magnets and Magnetism.

Nowadays the most common association with "magnets" is the colorful pieces every one sticks to a refrigerator. However the origin of this word comes from the name of a place in Asia Minor (modern Turkey): Magnesia. There loadstone or magnetic iron ore was found in ancient time. The ore contains magnetite, a natural magnetic material $F_3O_4$. Since ancient times it was noticed that a lodestone attracts objects made of iron and other lodestones. A legend tells about Archimedes successfully sinking enemy's warships by pulling iron nail out with the aid of a lodestone (which is probably not true). Socrates mentioned the ability of a lodestone to pass this property to an iron nail or ring after being in contact with the lodestone. Or as we say now a lodestone magnetizes an iron object.

The first practical application of magnets was the compass. No one knows who came up with the idea first Greeks, Arabs, or Chinese but it is documented that around 1000AD Chinese used magnetized needle floating on a reed in a bowl of water to find the south – north direction.
Columbus used a compass in his voyage to the Americas. He noticed that the northern direction pointed by a compass is slightly off from the true North as indicated by the stars. This observation was explained by a physician of Queen Elisabeth I, William Gilbert. He correctly suggested that Earth itself is a giant magnet and that the compass indicates the direction to the North magnetic pole which is different from the geographical North Pole.

For quite some time the phenomenon of magnetism was seen as different from the phenomenon of electricity but in 1821 Danish scientist Oersted noticed the influence of an electrical current on a magnetic needle. This was studied by French scientist Andre Ampere who concluded that the origin of magnetism is actually in ...current electricity. The magnetic force is nothing else but the force between electrical currents: parallel currents in the same directions attract but in opposite directions – repel. Therefore magnetic poles are just indicators of the direction of electrical currents. Ampere also was able to explain the magnetic properties of iron by assuming a circular current within each atom of iron. That current makes each atom of iron a magnet and when all atoms-magnets are aligned the entire piece of iron becomes a magnet itself.

There was still the question of how magnetic interaction could happen without apparent physical connection between magnets. Or is this connection present but one cannot see it?
The answer came from English scientist Michael Faraday who introduced the idea of electrical and magnetic fields. He came with a clever solution to illustrate these fields with the aid of lines of force, as he called them, which represent the force at each point of space occupied by the field. Through experiments Faraday found the connection between electrical and magnetic fields but he could not connect them mathematically. This was done by James Maxwell 50 years later. Maxwell tied together magnetic and electrical fields in four equations known as "Maxwell's equations" and showed that they are just two sides of the one electromagnetic field. Although it seems like "Maxwell's equations" finalized our knowledge about electromagnetism (and a lot of people thought so at the end of 19th century) in reality they just opened the door into totally different world – the world of modern physics.

# Chart 4.3: "It's All About Magnets..."

*Magnet:* object which attracts iron, cobalt, nickel, and neodymium

*Lodestone:* natural magnet

| Permanent | Temporary: while electrical current flows |

*Magnetism:* phenomena by which magnetic materials attract or repel

The nature of magnetism is in the *motion of electric charges.* Spinning and revolving electrons in an atom produce a magnetic field. Every spinning electron is a little magnet. If two electrons spin in opposite directions they cancel each other's magnetic fields. In iron or another magnetic material the fields are not canceled because of their specific atomic structure. Even more, individual atoms *line up with one another* in large clusters called ***Domains.***

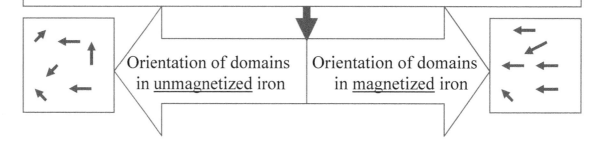

Orientation of domains in <u>unmagnetized</u> iron

Orientation of domains in <u>magnetized</u> iron

*Lines of Force:*
Represent the magnetic field and show the direction of the magnetic force.

*Magnetic Poles:*
Ends of magnets about which the force of attraction/repulsion seems to be concentrated.

- *Like poles repel*
- *Opposite poles attract*

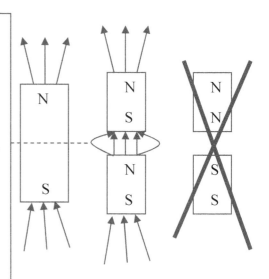

## Activity 4.5: "Fishing Expedition!"

Magnetism is usually presented along with electricity as it is based on charges as well. We have all experimented with magnets attracting and repelling each other and one might draw a parallel with static electricity phenomena. This would not be correct there are no such things as "south magnetic charges" or "north magnetic charges".

**Magnetism is the result of the movement of electrical charge.**

In the experiment below, you will give a "magnetic test" to a set of materials. First predict which materials will have magnetic properties. How many "fish" do you think you can catch? Then approach a magnet....

| Material Tested | Prediction (Would a magnet catch it?) | Result (Is it in your "fish" basket?) | Comment on magnetic properties |
|---|---|---|---|
| Chalk | | | |
| Aluminum | | | |
| Steel | | | |
| Brass | | | |
| Nylon | | | |
| Wood | | | |
| Copper | | | |
| Iron | | | |

List of Magnetic materials:

_____

# *Activity 4.6:* "Fishing Expedition - II"

Look at the pictures shown below. Circle the objects that a magnet will attract.

## Brass keys

© Robert Pernell, 2009. Under license
from Shutterstock, Inc.

## Wax candle

© sgame, 2009. Under license
from Shutterstock, Inc.

**Paper**

© granata1111, 2009. Under license
from Shutterstock, Inc.

## Paper clips

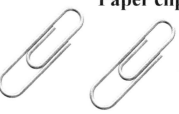

© Robert Spriggs, 2009. Under license
from Shutterstock, Inc.

## Credit card

© Thank You, 2009. Under
license from Shutterstock, Inc.

## Golden ring

© Elnur, 2009. Under license
from Shutterstock, Inc.

# 4.6 Waves Around Us

Look around and you will find a lot of examples of waves. Ocean waves, sound waves, and light waves are a few cases. How about a swing set? Is it an example of a wave? No, it is just a repetitive or periodic motion. In order to be called a wave the motion needs to travel but the swing set stays at the same location. Waves are caused by disturbances of different nature for the different waves: ocean waves by winds, earthquake waves by motion of tectonic plates, sound waves by the vibration of an object, and light waves by changes in an electromagnetic field. There are two general categories of waves: longitudinal waves when the periodic motion occurs along the traveling direction and transverse waves when the periodic motion occurs perpendicular to the traveling direction. Ocean waves are transverse. Traffic waves, when cars periodically bulk at the traffic lights or spread along the road in between, are longitudinal. Earth quakes create both transverse and longitudinal waves.

Light is a transverse wave. Since a sound wave is a result of compression and decompression of media that transmits the sound, it is a longitudinal wave.

**Exercise:**
Place the names from the petals of the Daisy in the proper box

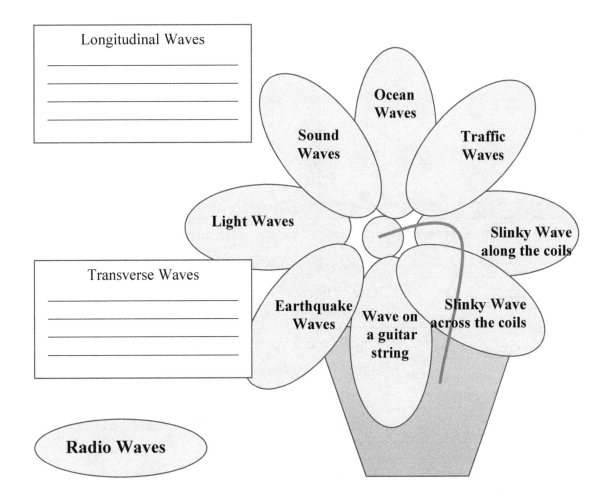

117

# Chart 4.3: "It's All About Waves…"

## Waves

A wave is a traveling disturbance carrying energy from place to place. There is no transfer of mass or material, the medium through which the wave travels goes back to its original state after the wave has passed.

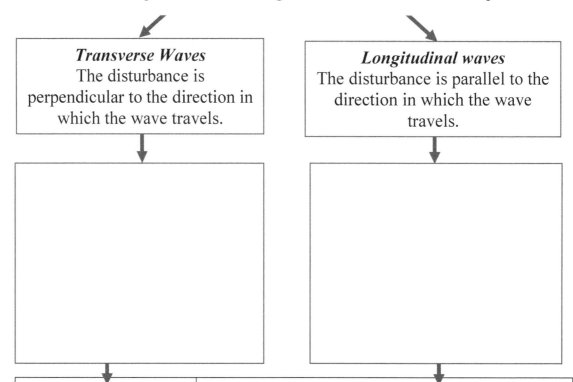

| Wave Characteristics | Definition |
|---|---|
| Wave Crest | The top of the wave (maximum of the disturbance) |
| Wave Trough | The bottom of the wave (minimum of the disturbance) |
| Wavelength | Distance between two consecutive crests or troughs |
| Frequency | The number of waves that pass per unit time |
| Amplitude | Half the distance between the crest and the trough of the wave |
| Wave Speed | The speed at which the wave travels |

# *Activity 4.7:* **Standing Waves**

When waves move back and forth along a string or within a tube for example they can sometimes combine to form a standing wave. The standing wave is created if the crests of incoming and reflected waves always combine at the same location.

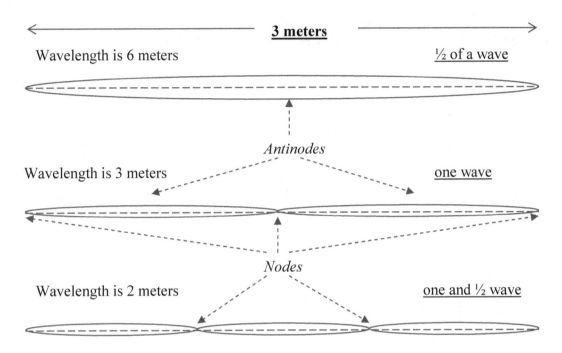

Each group of student will take one or two of the strings with a small electrical motor attached to the end of the string. Insert a battery in the electrical motor. Hang the motor at the end of the string and see how many types of standing waves you can generate. Once you generate standing waves measure the following:

| Exp. # | Number of Nodes | Number of Antinodes | Wavelength | Amplitude |
|--------|-----------------|---------------------|------------|-----------|
| 1      |                 |                     |            |           |
| 2      |                 |                     |            |           |
| 3      |                 |                     |            |           |

# 4.7 Sound

Sound is a longitudinal wave generated by a vibration of a material object. Because of that it requires a media for transmission and can not travel in vacuum. Sound travels with different speeds in different materials. It travels slower through gases than liquids or solids.

## Chart 4.4: "It's All About Sound..."

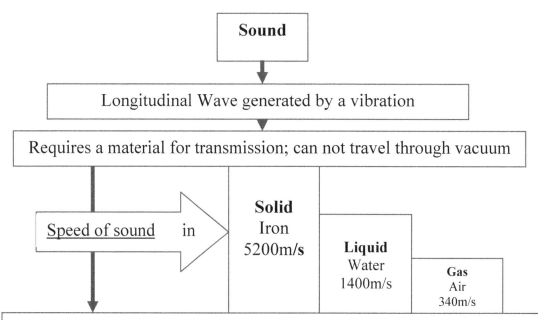

The range of human hearing is from 20 cps to 20000 cps (or Hz). Sound waves lower than 20Hz are called infrasound, and higher than 20kHz are called ultrasound. Humans are not able to hear ultrasound or infrasound. Other species such as dogs, bats and dolphins can hear ultrasounds, and elephants and whales can hear infrasound.

The lowest frequency of the human voice is about 100 Hz. A voice in range from 80Hz to 330Hz is called base and normally belongs to males. The highest frequency reached by a human voice is about 1400Hz. A voice of that level is called coloratura soprano and is usually associates with part of *Queen of the Night* from Mozart's *Magic Flute*.

***Activity 4.8:*** **Example of Activity Designed for Special Education**

# Sound

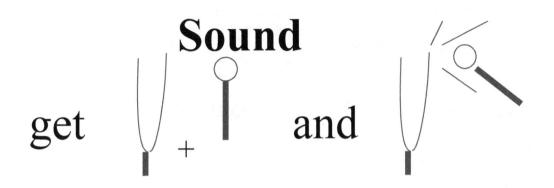

get     +     and

## What do you hear?

— — — — — — —

## What do you see?

— — — — — —

## What made the splash?

— — — — — — — — —

## What made the sound?

— — — — — — — — — — —

# 4.8 Light

Do you know that most light waves are not visible to the human eye?
Radio waves, microwaves, infrared waves, UV, X-rays, and γ-rays are all light waves. But these light waves are invisible to the human eye. The part of light that is visible to the human eye or visible light ranges from 400nm to 700nm of wavelength. Most birds and honey bees are able to see in the near UV range; snakes have a second set of "eyes", a special sensory organ that detects infrared light.

## Chart 4.5:  "It's All About Light…"

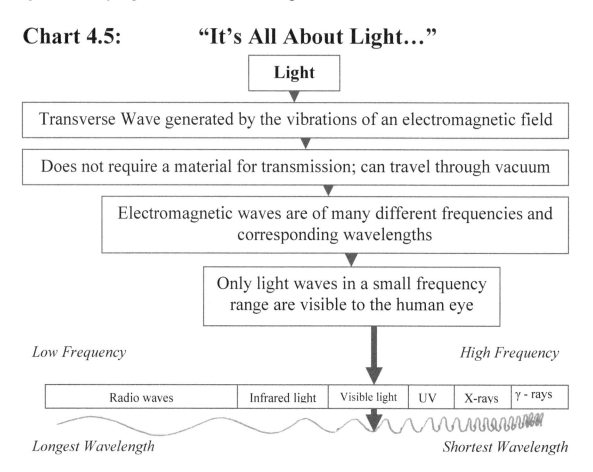

As the frequency of the light increases the wavelength gets shorter.
Short wavelength lights (UV, x-rays, γ – rays) are high energy lights
and therefore are hazardous.

Characteristics of visible light

- **Color** – corresponds to the frequency of a light wave
  - ❖ Visible light spectrum includes *red, orange, yellow, green, blue, and violet*
  - ❖ Red light has the longest wavelength and the lowest frequency
  - ❖ Violet light has the shortest wavelength and the highest frequency
- **Intensity** –depends on the amplitude of a light wave

## *Activity 4.9:* Sound vs. Light

Although both are waves sound and light are very different in nature, type, properties. To review your knowledge about sound and light complete the table below. Use this table to answer the following questions that require comparing the properties of sound and light.

|  | **Light Waves** | **Sound Waves** |
|---|---|---|
| Type of wave |  |  |
| How is it generated? |  |  |
| Are there any requirements for transition? |  |  |
| Where it can/can not travel |  |  |
| What is its speed in air? |  |  |

A) Why can we see but not hear violent outbursts on the surface of the Sun?

_____

_____

B) Why if you drop a rock into a well at first you see the splash and only later do you hear it?

_____

_____

C) Why sea mammals such as dolphins and whales have a good hearing but not a good vision?

_____

_____

# 4.9 Reflection and Refraction of Light

When a wave such as a light wave encounters an interface such as a change in material or light coming upon a mirror or light going from air to water the wave will be in part reflected and in part refracted.

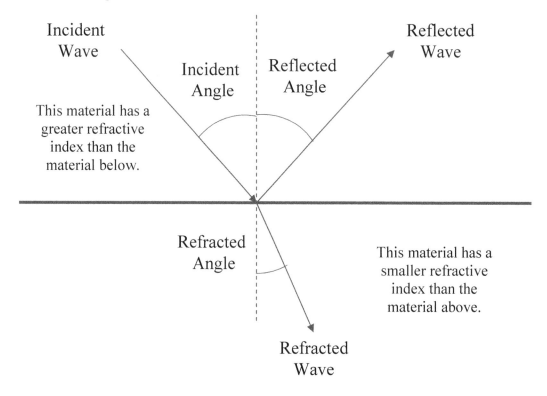

Figure 4.3. Reflection and Refraction of Light.

The incident, reflected and refracted angles are all measured with respect to a line perpendicular to the surface. The reflected angle is always the same as the incident angle. This relationship is called the Law of Reflection.

The refracted angle will depend on the difference in optical properties between the materials on the incident and the refracted sides. If the incident side is more optically dense than the refracted side (from glass to air) then the refracted angle will be greater than the incident angle and the light ray will be bended away from the perpendicular line. If the incident side is less optically dense than the refracted side (from air to water) then the refracted angle will be less than the incident angle and the light ray will be bended toward the perpendicular line. This relationship is called the Law of Refraction and is described by the refractive indexes of the materials. The greater the refractive index the greater the optical density of the material.

Light of different frequencies have different indexes of refraction for the same material. Lights of different colors refract or bend at different angles in the same materials. As a result white light can be separated into its components and we see a full spectrum or a rainbow. Refraction is behind many natural optical phenomena such as a mirage.

# *Activity 4.10:* **Mirrors, Lenses and Prisms**

In the case of light waves, one can use surfaces to make mirrors and the change of material from the incident side to the refracted side can be used to make magnifying glasses or prisms (there are other examples/uses).
The ray box will help you to understand how mirror and lens work.

a) Put one slit plate into the ray box and place it on white paper. Then direct the light ray coming from the ray box <u>to the surface of the mirror</u> at a certain angle. You should see how the ray coming from the box is reflected by the mirror. Now take a pencil and draw a line along the edge of the mirror. To trace both rays (incident and reflected) mark two dots inside of each ray then connect the corresponding dots and extend the lines.

b) Now examine the trace picture. Find a point where the incident ray becomes reflected and draw a line perpendicular to the mirror surface from that point. Fold the paper along this line then fold the paper two more times first along the incident ray and second along the reflected ray. Have you noticed something about the incident and reflected angles while folding? Explain.

_____

_____

_____

_____

_____

c) Put a 5 slits plate into the ray box. Place the ray box back on the white paper to see the five parallel rays. Take both lenses from the plastic box and place each in order across the rays. You will see that lenses of different shapes cause different effects on parallel light. The one which gathers light rays together is called a **convex lens**. The other type is called a **concave lens**. For each lens trace the ray's path.

Which one of them is used in magnified glass? What precautions should you take with this lens? Why?

_____

_____

_____

d) In the case of a prism the light rays of the different colors are refracted differently (different refractive indexes) when the light goes into the prism and then exit the prism. The result: at the exit of the prism the light ray is split and one can see the different colors of the incident white light. Put one slit plate into the ray box and place it on the white paper. Then direct the light ray coming from the ray box to the side of a prism. Trace the prism and rays coming to the prism and exiting the prism. Label colors of rays exiting the prism in accordingly.

What is the order of the colors from the most bended to the least bended?

_____

# Topic 5

# The Earth and Beyond

# 5.1 It's All About the Earth...

Earth is the third planet of a the planetary system of the star Sun. Our home is located in the outer regions of the Milky Way. The Milky Way is one of many billions of galaxies in our universe. The earth is our home, a small iron rich rocky planet with an atmosphere and life. The conditions on all the solar system objects are very different and none other than earth can sustain life. For most of human history we were just living on the earth not knowing much about the rest of the solar system and the universe. Last century saw an explosion of knowledge about the space around us; we even walked on our moon, sent probes to other planets and placed telescopes in orbit. But how much do we know about our Earth?

During the early part of human history our impact on the planet was very modest and localized. As humanity expanded and particularly since the industrial revolution mankind is increasingly affecting the functioning of the Earth. Last century we were able to solve or mitigate big environmental problems such as a hole in our ozone layer or lead in our atmosphere. Mankind is presently facing its biggest environmental challenge with the onset of a rapid warming likely caused by green house gases such as carbon dioxide and methane. These gases are a byproduct of our way of life and substantial changes will be required to avoid profound changes such as changes in local climatology affecting ecosystems and agriculture, substantial sea level rises etc. We need to continue to study our earth and educate as large a portion of the population as possible to make the best decisions. We will need to make tough decisions to balance economic imperatives while preserving a climate favorable to human life. Once again will need to count on human ingeniosity to develop solutions and it will start with the continuously improving scientific understanding of our earth system.

Figure 5.1. Photo of the earth in space © Ali Ender Birer, 2009. Under license from Shutterstock, Inc.

# Chart 5.1:     "Earth's Spheres"

To study our planet scientists divide the earth into four components or spheres. These spheres all interact with each other. Chart 5.1 illustrates the four spheres: the ***Geosphere*** the ***Hydrosphere,*** the ***Atmosphere***, and the ***Biosphere***.

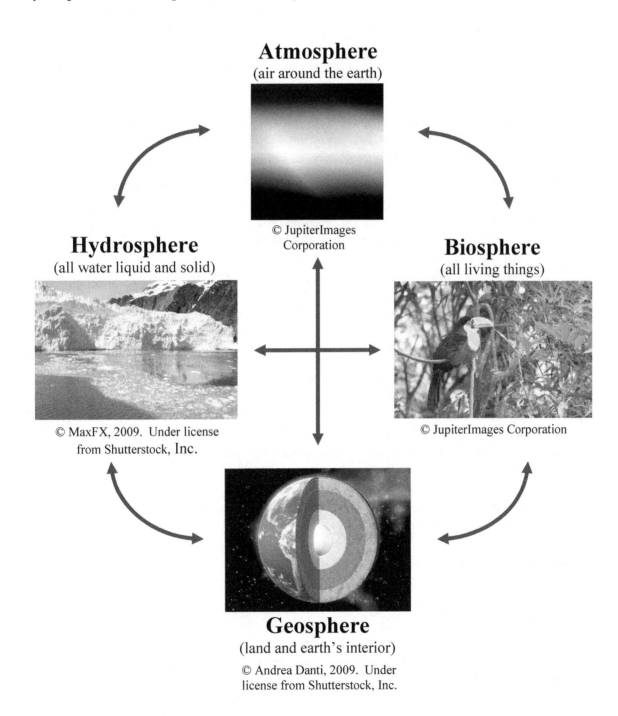

**Atmosphere**
(air around the earth)

© JupiterImages
Corporation

**Hydrosphere**
(all water liquid and solid)

© MaxFX, 2009.  Under license
from Shutterstock, Inc.

**Biosphere**
(all living things)

© JupiterImages Corporation

**Geosphere**
(land and earth's interior)

© Andrea Danti, 2009.  Under
license from Shutterstock, Inc.

**The Geosphere:** *consists of layers of materials, solid and liquid, starting at the surface (crust) of our earth and down to its center. The layers are physically and chemically different.*

- The crust consists of the rocks and the soil materials. It is rich in nutrients, oxygen, and silicon.
- The mantle is the thick semi-solid land underneath the crust. It is composed of oxygen, silicon, iron and magnesium.
- The core of the earth consists of metallic liquid materials and then solid materials at the very the center of the planet, both the liquid and solid portions of the core are made of iron and nickel.

**The Hydrosphere:** *consists of all the solid, liquid, and gaseous water of our planet starting from below the bottom of the oceans in the lithosphere all the way high into the atmosphere.*

- Most of the hydrosphere consists of the salty water of our planet's oceans
- A small portion of the water in the hydrosphere is fresh (non-salty).
- Fresh water evaporates and then flows as precipitation from the atmosphere down to Earth's surface, as rivers and streams along the Earth's surface, and as groundwater beneath Earth's surface.
- Most of Earth's fresh water is frozen.

**The Atmosphere:** *consists of all the air in Earth's system starting in the ground at the surface of the geosphere and extending up to more than 10,000 km above our planet's surface.*

- The atmosphere is divided in several distinct layers with air temperature alternatively rising and decreasing in the layers.
- The upper portion of the atmosphere protects the biosphere from the harmful high energy portions of our sun's rays, ultraviolet radiation.
- The atmosphere traps heat which makes the temperatures on earth livable for humans.
- Weather occurs in the bottom layer of the atmosphere. As air warms up and cool down it rises and sinks and moves around the planet resulting in anything from light breezes to devastating hurricanes.

**The Biosphere:** *consists of all the planet's living things. Besides human and animals this includes all of the microorganisms and the plants of the Earth.*

- Within the biosphere, living things form ecological communities based on the physical surroundings of an area.
- These communities are referred to as biomes. Deserts, grasslands, and tropical rainforests are three of the many types of biomes that exist within the biosphere.
- The biosphere interacts with all the other spheres, changing the geosphere surface, constantly modifying and regenerating our atmosphere, sustaining itself from our hydrosphere.

There is one important component missing from Chart 5.1. For the earth to function we need the energy of our sun and we could add the Heliosphere to the chart but we will not study the sun in any details here. Some of the spheres in Chart 5.1 can be further divided into smaller components such as the cryosphere (the frozen water).

**Exercise:** Find examples for each of the earth spheres (e.g. animals in the biosphere…).

**The Geosphere:**   1. _____    2. _____

                            3. _____    4. _____

**The Hydrosphere:**   1. _____    2. _____

                            3. _____    4. _____

**The Atmosphere:**   1. _____    2. _____

                            3. _____    4. _____

**The Biosphere:**   1. _____    2. _____

                            3. _____    4. _____

**Exercise:** The interactions between spheres are just as important as the spheres themselves. Find examples of processes that illustrate the interaction between spheres, e.g. wind erosion is an interaction between the atmosphere and the geosphere.

**Biosphere - Atmosphere:**

1. _____

2. _____

3. _____

**Geosphere - Atmosphere:**

1. _____

2. _____

3. _____

**The Hydrosphere - Biosphere:**

1. _____

2. _____

3. _____

## *Activity 5.1:* "An Interacting System"

The interaction between the spheres takes place in different ways but also following different time scales. Some interactions take place very quickly (e.g. changes due to weather), others take million of years (e.g. changes due to plate tectonics). The graphic below illustrates some of these processes.

Figure 5.2. Interactive processes for a small bay. © JupiterImages Corporation

Look at the picture above and list for your area the relevant interactive processes.

List at least three natural events that can greatly affect the balance of the environmental system in your area.

(a) _____     (b)_____     (c)_____

# 5.2 The Atmosphere

The atmosphere contains all the air that surrounds our planet. Without our atmosphere we would not have life as we know it. The atmosphere controls also everyday weather and on a long term range along with the other spheres the climate.

## Chart 5.2: "Our Atmosphere"

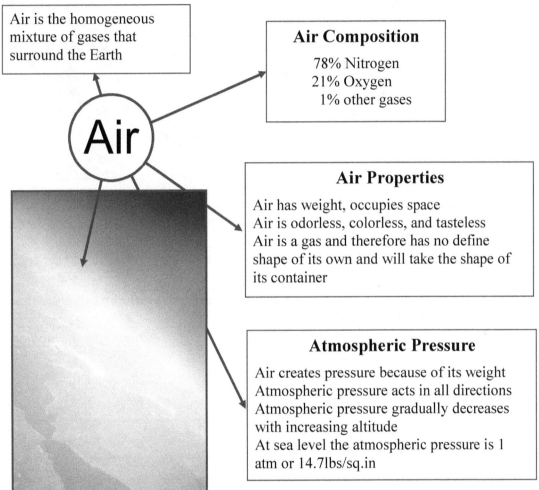

Air is the homogeneous mixture of gases that surround the Earth

**Air Composition**

78% Nitrogen
21% Oxygen
1% other gases

**Air Properties**

Air has weight, occupies space
Air is odorless, colorless, and tasteless
Air is a gas and therefore has no define shape of its own and will take the shape of its container

**Atmospheric Pressure**

Air creates pressure because of its weight
Atmospheric pressure acts in all directions
Atmospheric pressure gradually decreases with increasing altitude
At sea level the atmospheric pressure is 1 atm or 14.7lbs/sq.in

Figure 5.3. Atmosphere above the earth. © JupiterImages Corporation

**Exercise 5.3:**
What are the main gases we inhale?

_____

What are the main gases we exhale?

_____

## Activity 5.2: "The Imploding Soda Can"

A classic demonstration to illustrate the importance of atmospheric pressure goes as follows: take an empty aluminum can of your choice (any soft drink will do) - fill it with some water up to 10%-20% of the can - use metal tongs or other means to hold the can above a heat source and bring the water in the can to its boiling point. Once the water is boiling take the can with the tongs and quickly place it upside down in a bath of cold water and observe what happens.

Once the water is boiling what is its temperature?

_____

If we allow all the water to boil off, what will happen to the temperature of the can thereafter?

_____

(a)    Cold Water    (b)

Figure 5.4. Soda can (a) before and (b) after being crushed by the pressure outside the can.

In the table below check the appropriate box to indicate if the air pressure inside the can is the *lower*, the *same*, or *higher* than the air pressure in the room.

| Steps of the Demonstration | The air pressure inside the can is ... than in the room | | |
|---|---|---|---|
| | *lower* | *same* | *higher* |
| At the beginning of the demonstration before the can is heated | | | |
| When the water is boiling in the can | | | |
| After the can opening is sealed and the can is cooled when placed in the water | | | |

Explain why the soda can is crushed once it is placed upside down in the cold water

_____

What would happen to the soda can if it is placed with its opening facing upwards rather than facing down in the water? Crushed or not crushed? Why?

_____

What would happen to the soda can if we let all the water boil off before placing it upside down in the water?

_____

134

# Chart 5.3: "Layers of the Atmosphere"

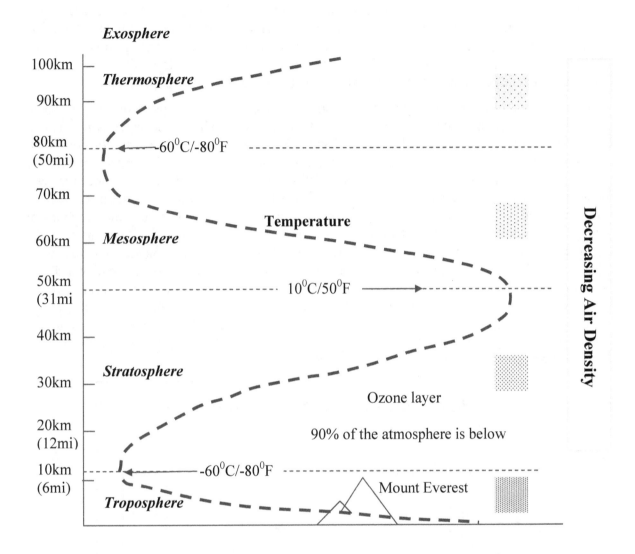

- ✓ We live in the lower layer of the atmosphere
- ✓ Clouds can be found **only** in the Troposphere
- ✓ The Troposphere is also the weather layer
- ✓ The ozone layer is part of the stratosphere
- ✓ The atmosphere protects us space high energy radiation
- ✓ 99% of air is in the troposphere and stratosphere
- ✓ The exosphere is the merging layer

# (Web) Activity 5.3: "Good Ozone – Bad Ozone"

Ozone is a molecule made of 3 oxygen atoms, $O_3$. Small amounts of ozone are found throughout the atmosphere but most of the ozone is found in two locations. About 10% of the ozone is found at ground level it is also referred to as "smog" ozone. Ozone on the ground is a pollutant with toxic impact on humans and vegetation. When ozone is predicted to be high, cities enact ozone action days to try to lower ozone levels and warn the population.

Most of the rest of the ozone (~ 90%) is found much higher in the atmosphere between about 10 and 50 km above ground, it is called atmospheric ozone. The ozone layer is referred to where most of that ozone resides between 20 and 25km above ground. The vibrations of the $O_3$ molecule allow it to absorb UV light rays decreasing substantially the UV rays reaching the earth and protecting us from this dangerous radiation. In 1981-83 a hole in our ozone layer was detected in Antarctica. Researchers found out that the hole was growing and that it was related to the presence of chlorine coming from man made chemicals such as chlorofluorocarbons or CFCs, broken down by UV rays. CFCs were a great invention early in the 20th century to make refrigeration safer. It turned out however that these and other similar man made chemicals were also slowly rising in the atmosphere and leading to the destruction of atmospheric ozone. Three scientists were awarded the 1995 Chemistry Nobel Prize for their work on the formation and decomposition of ozone. While some CFCs related regulations were enacted starting in the 1980s, it took the discovery of the rapidly growing ozone hole for governments to act decisively and ban the production of some CFCs starting in 1996. The ozone layer has now stabilized and we are hoping the recovery has started.

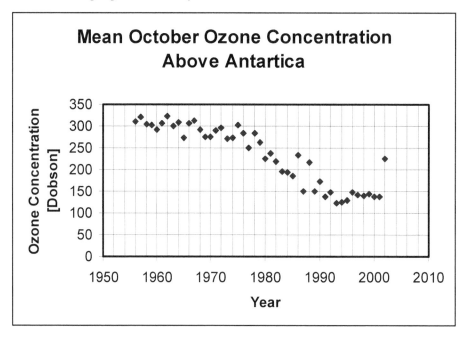

Figure 5.5. Mean October values of the ozone concentration at the Halley Research station in Antarctica. The data are collected using the Dobson spectrophotometer and are available from the US EPA website (http://www.epa.gov/spdpublc/science/indicat/techsupp.html).

(1) In which layer do we find atmospheric ozone?_____

(2) Log on to the US EPA website and find out the forecasted ozone levels for today and tomorrow for your location.

(3) Find one nearby location where the ozone level is predicted to be good or moderate and one location where the ozone level is predicted to be unhealthy or even dangerous. List differences between the locations that could explain the difference in predicted ozone levels?

(4) Log on to the US EPA website and find out the UV Index Forecast for your location for the next 4 days.

(5) What can you do to protect yourself from the UVs if needed?

(6) How are UV radiation related to the molecule $O_3$?

(7) Find the status of the Ozone hole above our planet and list the website. Is the recovery continuing?

(8) When is the ozone layer predicted to be back to normal, i.e. to pre 1960's levels?

(9) Log on to the US government www.airnow.gov website and identify and list other air contaminants that affect at times our air quality.

# 5.3 The Hydrosphere

The hydrosphere contains all the water of our planet, fresh and salty, solid, liquid and gaseous. The hydrosphere extends downward several kilometers into the Geosphere and upward up to the bottom of the stratosphere. The oceans cover 71% of our planet's surface. While freshwater is an essential ingredient for life most of our water is salty (~97%) making it unfit to drink and irrigate. Most of the freshwater (~78%) is locked up in ice, glaciers and polar caps. For the rest of the freshwater about 21% can be found in the sediments in the ground: groundwater. Many of our cities and human activities depend on groundwater supply. Less than 1% of the freshwater is found in streams, rivers and lakes and the amount of water vapor in the atmosphere is minuscule.

## *Activity 5.4:* **"Where is the Water?"**

Where can one find:       fresh water?_____

                                 salt water?_____

Where can we find water in:   liquid form?_____

                                 solid form?_____

                                 gaseous form?_____

Draw a pie chart dividing first earth's salty and fresh water and earth's fresh water into their components

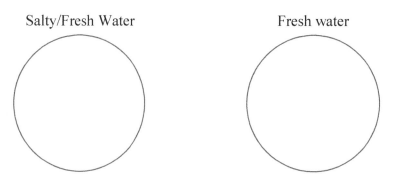

Salty/Fresh Water                    Fresh water

In what form is most of Earth's fresh water?

What can endanger sources of fresh water?

                           1._____

                           2._____

                           3._____

Can you name an aquifer (storing groundwater) near where you live?

How can all of us help keep our freshwater clean?

# Chart 5.4:                 "The Water Cycle"

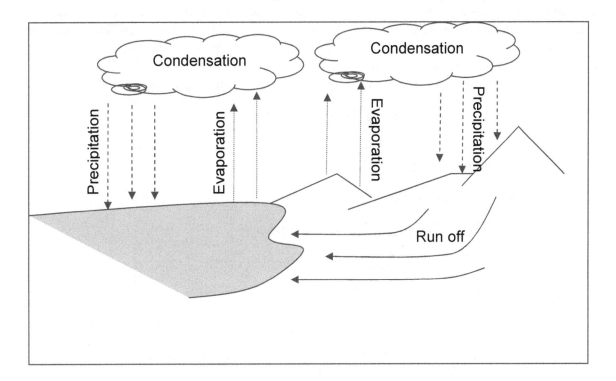

- The heat of the sun and gravity make water on Earth circulate all the time. *Evaporation* moves water from surface to the atmosphere where it condenses (*condensation*) and falls back to the surface as rain or snow (*precipitation*).

- If rain or snow falls in the ocean the cycle is complete. But if it falls on the ground, water drains to rivers or lakes or even into the ground where it continues its journey to the oceans all driven by gravity (*run off*).

- The natural circulation of water – from oceans to the atmosphere, from atmosphere to the ground and back to oceans – is the *WATER CYCLE*.

**Exercise 5.4:**
If evaporation and precipitation were not balanced what would happen?

More evaporation than precipitation:

_____

More precipitation than evaporation:

_____

The water that drains into the ground and fills the open pore space between sediments and rocks is called **ground water**. The region where water has completely filled all open pores is called the saturated zone. Above the saturated zone the portion of the soil up to the surface is called the unsaturated zone and contains various levels of soil moisture. Different surface materials drain water differently. For example sand easily soaks water up but clay does not. For a good aquifer one needs a soil that can contain a lot of water (high porosity soil) and below it a type of soil or rock that keep the water from draining deeper in the ground.

**Water Residence Time**

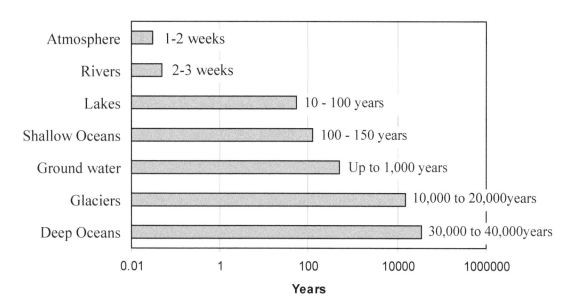

Figure 5.6. Residence time for a water droplet when it enters the atmosphere, enters a river, a lake etc…

# *Activity 5.5:* **"Water Drainage"**

**Objective:** The students will discover how water drains through different types of soil.

**Materials:** 4 cups with holes, 1 cup of water, 1 empty cup, 1 measuring cylinder, sand, clay, small pebbles, and medium rocks.

1. Place the same volume of sand, clay, small pebbles and medium rocks in each of the four cups with small holes at the bottom.
2. Measure 50ml of water and pour the water successively in each of the cups with materials while holding an empty cup below.
3. For each case, record the time that it takes for water to drain through the materials and measure the amount of water that went through the materials.
4. Build a chart with the drainage time and amounts of water drained for each material.
5. Repeat the experiment by testing different combinations of two types of materials e.g. small pebbles on top of a clay layer. Which combination will be favorable for an aquifer?

# (Web) Activity 5.6: "Currents & Climate"

Besides being essential for life water is also very important to regulate our climate. The heat capacity of water is large, i.e. it takes a lot of energy to heat or cool water. As a result water temperature varies much less than air temperature. Areas near the oceans have cooler summers and milder winters than areas away from the coast.

The oceans are not still but water flows in large oceanic currents. For example currents moving from the equator to the poles bring warmer water to cooler region. Log on to the world wide web to the NASA "Ocean Motion and Surface Currents" website at 'http://oceanmotion.org/'.

Find the map outlining the main ocean surface currents by testing the links at the top of the website home.

1. Find the Kuroshio, the Gulf Stream and the Brazil currents. How are these currents similar and different?

2. Find the Canary, the Peru and the California Currents. What is the main difference between these currents and the currents in question 1?

3. Which currents bring cold fresher water to the North Atlantic Ocean?

4. Which ocean current is responsible for the warmer climate in Rome, Italy than New York (both cities are located at the same latitude and therefore receive the same amount of sunlight throughout the year)?

By browsing other portion of the same website, answer the following questions:

5. What is the ocean conveyor belt and what does this conveyor belt transport across the oceans?

6. How did Benjamin Franklin figure out a good navigational route to sail from the Americas to Europe?

Finally test your knowledge by completing the website quizzes.

# 5.4 The Geosphere

In these notes we consider the geosphere to contain all of the cold, hard solid land of the planet's crust (surface), the semi-solid land underneath the crust or mantle, and the liquid land near the center of the planet. The term lithosphere is sometimes used to describe the same concept or just for the crust and the upper part of the mantle. The information for the structure of the earth and plate tectonics was developed based on the US Geological Survey (USGS) web presentation "This Dynamic Earth: The story of Plate Tectonics" by W. Jacquelyne Kious and Robert I. Tilling. The document can be found on the web at: http://pubs.usgs.gov/publications/text/dynamic.html.

**A few facts about the earth**

***Travel to the center of the earth:*** as one "travels" to the center of the earth the temperature increases, the pressure increases and the materials become denser (note: it is not possible to actually travel to the center of the earth without melting away). The increase in pressure is due to weight of the portion of the earth above.

***What is the earth made of?*** The mantle is made of silicate $(SiO_4)^{-4}$ rocks rich in iron, magnesium, and calcium. The core of the earth is mostly made of iron and nickel. The crust of the earth is made of wider variety of materials. Overall the earth is made of ~33% Iron, 30% Oxygen, 16% Silicon, 14% Magnesium, 2% Nickel and 5% other materials.

***Where does the Earth's magnetic field come from?*** The core of the earth is composed of two layers, a liquid outer core and a solid inner core. As the earth rotates, the liquid of the outer core spins, which creates the earth's magnetic field.

**Two types of crusts:** the ***continental crust*** is between 20 and 60 km thick. It is composed of granite rocks, which are less dense than the basaltic rocks of the ***oceanic crust***. So, most of continental crust is above sea level. The **oceanic crust** is only about 10 km thick and is composed of basaltic rocks. These rocks are denser than the granite rocks of the continental crust. So, the oceanic crust is below sea level.

## *Activity5.7:    "The Earth and a Boiled Egg?"*

Boil and egg (about 7 minutes) and once boiled and cooled, cut the egg in half. Identify the component of the boiled egg with the following components of the earth structure:

| Earth's Component | Boiled Egg Parts |
|---|---|
| Earth's thin solid crust | |
| Earth plastic mantle | |
| Earth core | |

## Activity 5.8: "What's Beneath our Feet?"

A. Look at the diagram of the earth below and answer the questions.

**Crust:** 0-100km thick (solid)

**Mantle:** 100-2,900km (semi-solid)

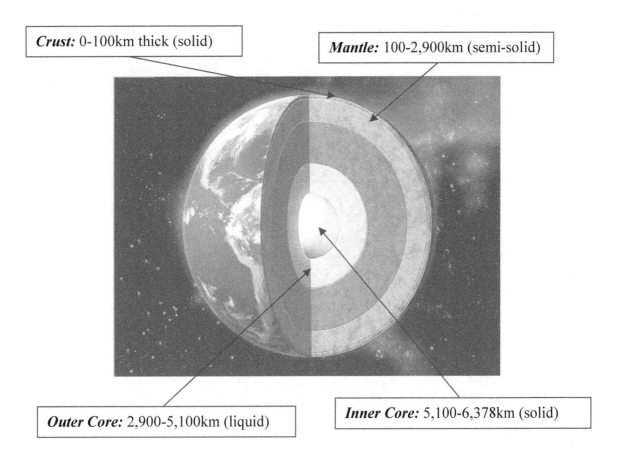

**Outer Core:** 2,900-5,100km (liquid)

**Inner Core:** 5,100-6,378km (solid)

Figure 5.7. Cutaway of the earth's interior. © Andrea Danti, 2009. Under license from Shutterstock, Inc.

What is the only liquid layer of the earth?_____

What is the thicker of the following layer or sub layers: the inner core, the outer core, the mantle or the crust:_____

What is the thinner of the following layers and sub layers: the inner core, the outer core, the mantle or the crust:_____

Which layer are the Rocky Mountains part of: the inner core, the outer core, the mantle or the crust?_____

B. Use sweet materials to construct the layers of the Earth.

1. Hershey's "Kiss" candy
2. Chocolate syrup
3. Chocolate ice cream
4. Caramel hard shell syrup
5. Ice cream sugar cone

Before starting you need to decide what layer of the Earth each sweet material represents. Match each of the sweet materials with the appropriate layer on the diagram below based on the similarity in its state of matter.

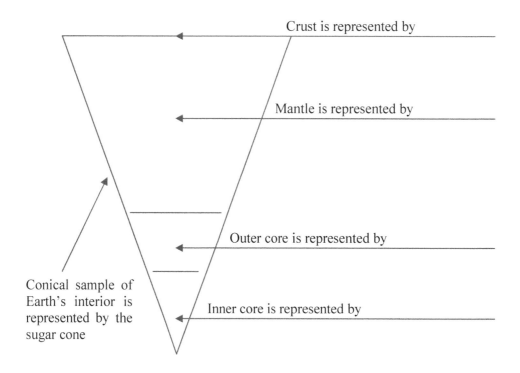

Fill up the cone with sweet materials accordingly to the diagram. Pay a special attention to the amount of chocolate syrup, chocolate ice cream, and caramel hard shell syrup. Keep in mind what <u>the mantel is the thickest layer and the crust is the thinnest</u>.

Wait till hard shell syrup hardens enough and poke a hole through the shell; then carefully press the shell down with your fingers to create a "volcano eruption".

Look at the choice of sweet materials: the sweets representing both parts of the Earth core are pure chocolate and the sweet representing the mantle is partially chocolate but the sweet representing the crust has no chocolate component at all.

If the chocolate represents iron what can you tell about the distribution of iron inside the Earth?
_____
_____

144

**Chart 5.5:**          **"Earth Structure"**

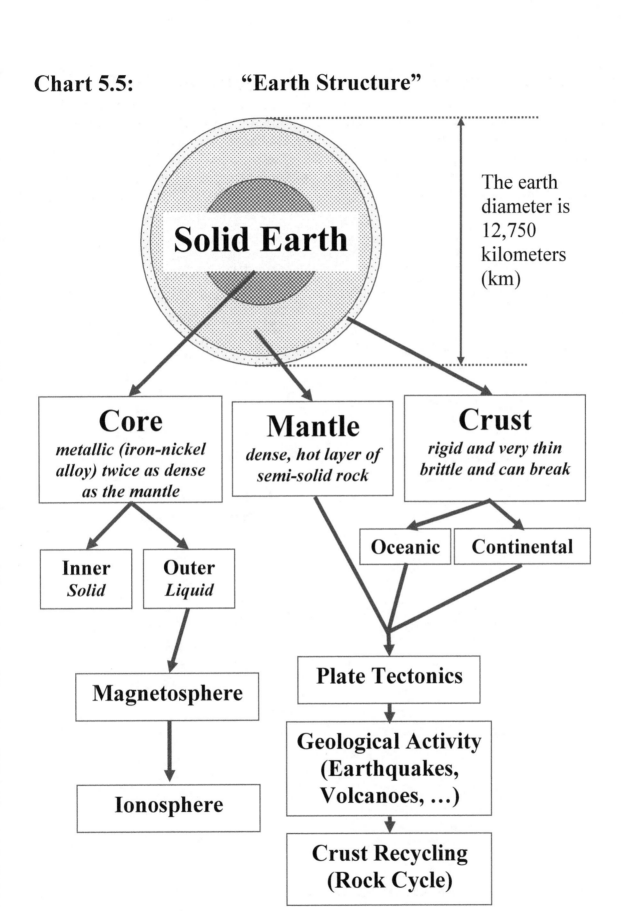

The earth diameter is 12,750 kilometers (km)

**Solid Earth**

**Core**
*metallic (iron-nickel alloy) twice as dense as the mantle*

**Mantle**
*dense, hot layer of semi-solid rock*

**Crust**
*rigid and very thin brittle and can break*

**Inner**
*Solid*

**Outer**
*Liquid*

**Oceanic**

**Continental**

**Magnetosphere**

**Ionosphere**

**Plate Tectonics**

**Geological Activity (Earthquakes, Volcanoes, ...)**

**Crust Recycling (Rock Cycle)**

# Chart 5.6:    "Plate Tectonics and Continental Drift"

By studying rock formations, fossil distributions across continents and the motion of the plates, scientists built the theory of continental drift. About 200 to 225 million years ago there was only one large continent on the earth: Pangea. The continued motion of the plate tectonics split this single continent and progressively moved the plates to today's configuration. Of course plate tectonic motion continues and in several million years, the earth will look different. (graphics courtesy of USGS).

### PERMIAN
225 million years ago

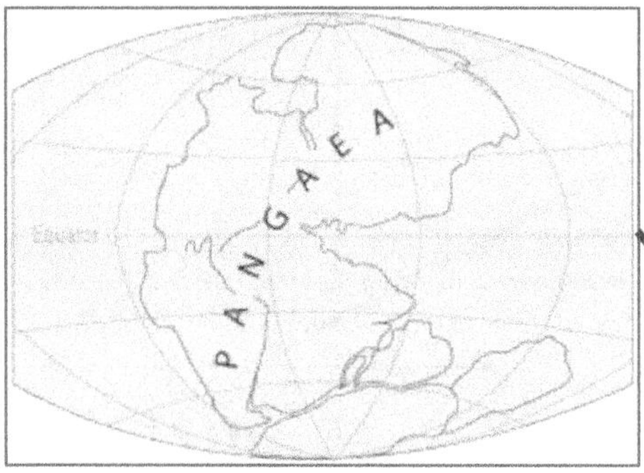

### TRIASSIC
200 million years ago

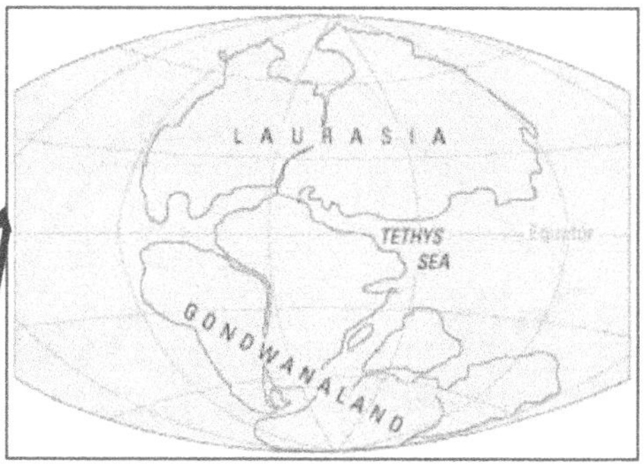

### JURASSIC
150 million years ago

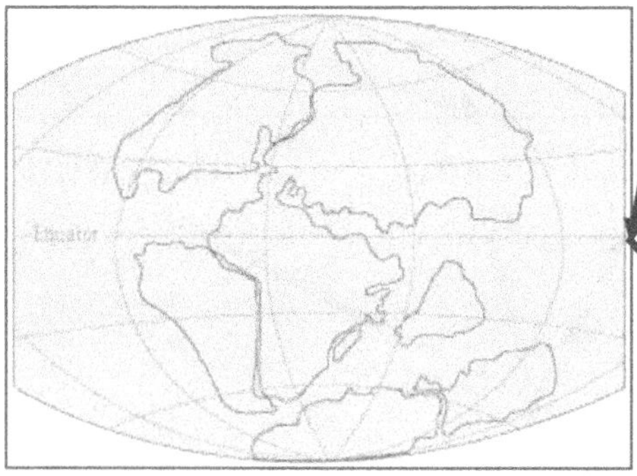

### CRETACEOUS
65 million years ago

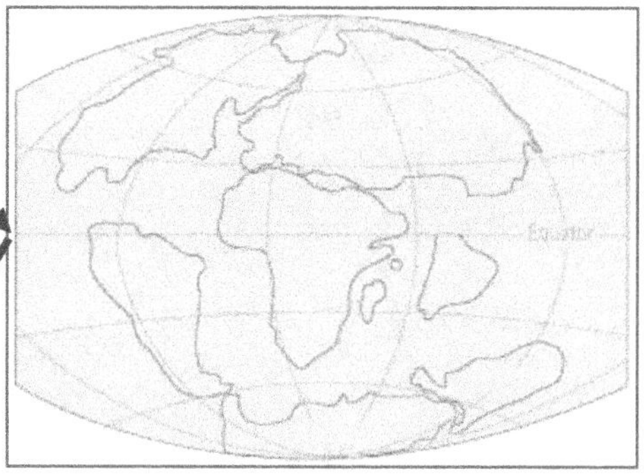

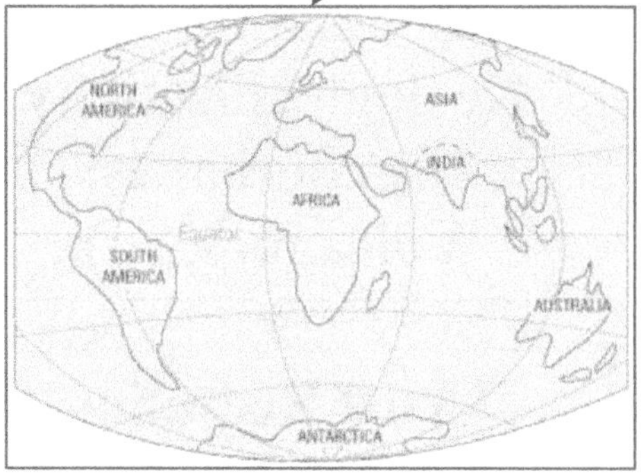

### PRESENT DAY

## Activity 5.9: "Tectonic Plates"

According to the theory of plate tectonics, the earth's crust is composed of large plates which contain all the world's continents and oceans. The plates' sizes vary greatly, from a few hundred to thousands of kilometers across. Most of the boundaries between individual plates cannot be seen, because they are hidden beneath the oceans. A well known exception is the San Andreas Fault in California. The plates are "floating" on the upper part of the mantle. The plates probably formed very early in the earth's history and have been drifting ever since. An example is provided in Figure 5.8 (a) with the motion of the Indian plate moving towards and colliding with the Eurasian Plate. Geologic activity such as volcanoes or earth quakes will take place where plates collide. An illustration of the present arrangement of the tectonic plates is presented in Figure 5.8 (b). Based on these illustrations answer the following questions.

What land formation resulted from the collision between the Indian and Eurasian Plate?_____

Which tectonic plates surround the Coco's plate?:_____

The relative movement of which plates is responsible for earth quakes in San Francisco?
_____

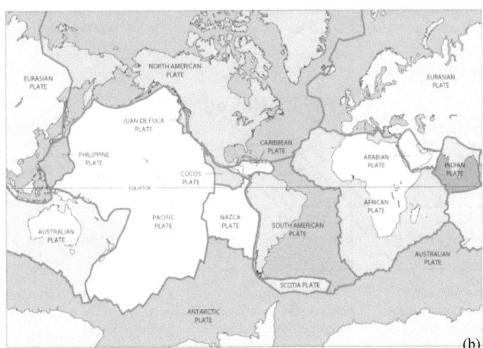

Figure 5.8: (a) Present day location of tectonic plates and (b) motion of the Indian plate. Graphics courtesy of USGS.

## *Activity 5.10:* "Plate Tectonics and Geological Activity"

Before the theory of plate tectonic, scientists had noticed that earth quakes were only taking place in very narrow zones. Today information about recent earthquakes is automatically published on the World Wide Web.
Log on to "http://earthquake.usgs.gov" and answer the questions below.

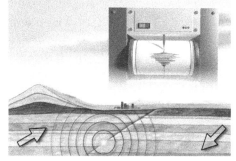

How many earth quakes did take place on earth over the last hour?_____

Figure 5.9. Illustration of Earth quake recorded at plate boundary.
© Andrea Danti, 2009. Under license from Shutterstock. Inc.

Where did the largest earth quake take place today and what was its magnitude?
_____

Find the earthquake closest to your location. When did it take place and what was its magnitude?
_____

There are usually groups of earthquakes. Between which tectonic plates did the largest group of earthquake take place during the past week?
_____

From the "About Earthquake" section of the website select "Animations for Earthquakes". Find the shadow zone animation. How does the shadow zone reveals that our earth has a liquid core?_____
_____

## Chart 5.7:            "Plate Boundaries"

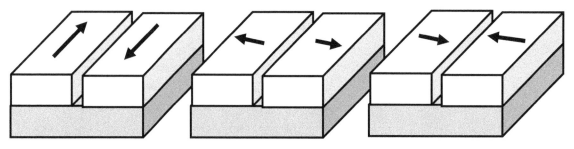

***Transform boundary:*** plates slide horizontally, crust is neither produced nor destroyed

***Divergent boundary:*** new crust is generated as plates pull away from each other

***Convergent boundary:*** crust is destroyed as one plate dives under another plate

***Plate boundary zones:*** broad zones in which the boundaries are not well defined and there are no clear effects on the plates

# Chart 5.8: "Rocks and Minerals"

**MINERALS** are the building blocks of rocks. The building blocks of the minerals are the elements in the periodic table.

A *mineral* is defined as ***a naturally formed, inorganic, crystalline solid***, composed of an ordered arrangement of atoms with specific chemical composition.

Of the known 112 elements, 92 occur naturally in the earth's crust and combine to make 4000 different minerals.

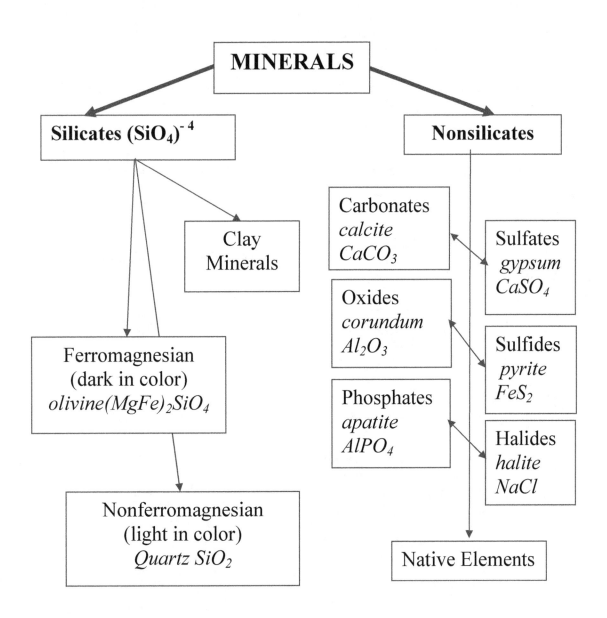

# Chart 5.9: "Types of Rocks and the Rock Cycle"

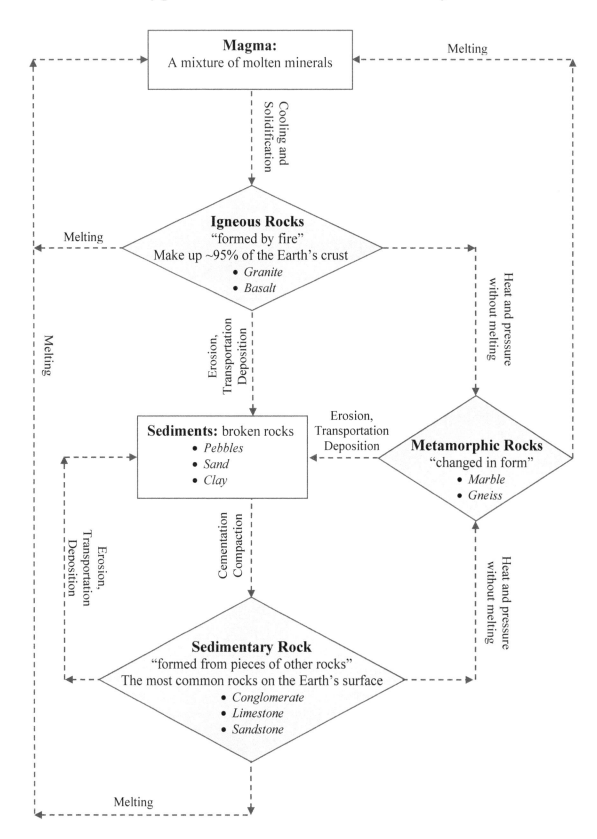

**Magma:**
A mixture of molten minerals

Cooling and Solidification

Melting

**Igneous Rocks**
"formed by fire"
Make up ~95% of the Earth's crust
- *Granite*
- *Basalt*

Melting

Erosion, Transportation Deposition

Heat and pressure without melting

**Sediments:** broken rocks
- *Pebbles*
- *Sand*
- *Clay*

Erosion, Transportation Deposition

**Metamorphic Rocks**
"changed in form"
- *Marble*
- *Gneiss*

Melting

Erosion, Transportation Deposition

Cementation Compaction

Heat and pressure without melting

**Sedimentary Rock**
"formed from pieces of other rocks"
The most common rocks on the Earth's surface
- *Conglomerate*
- *Limestone*
- *Sandstone*

Melting

# Activity 5.11: "Rock Cycle Dance"

**Objective:** To familiarize the students with the rock cycle and identification of the different types of rocks. Target grades: 4-5.

**Materials:**
- Various samples of igneous, sedimentary, metamorphic rocks and sediments (in cups)
- Five signs with the labels igneous, sedimentary, metamorphic rocks, magma and sediments
- A round table and masking tape

**Procedure:** With the strips of masking tape divide the table into five sectors. Place a corresponding sign in each sector; then place all the samples in the middle of the table as shown on the diagram below.

Ask five students to take a place around the table – one student per sector.

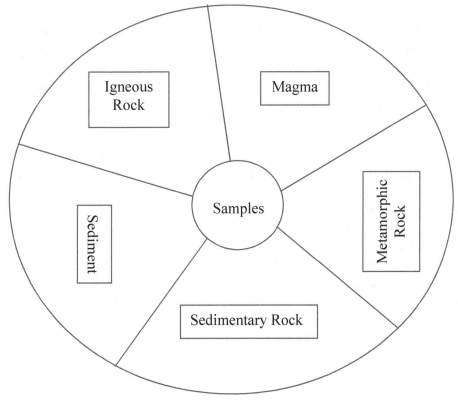

When the music starts, the students walk around the table turning clockwise. When the music stops each of the students must take a sample from the center corresponding to its present position (the student in the igneous sector of the table, must take an igneous rock, the student in the sediment sector of the table, must take a cup of sand or clay, etc.). The student in the magma sector does not take anything (it would be too hot for a person to touch without very specialized equipment).

After the samples have been collected, the other students from the class discuss the choices of rocks and sediments. If one of the choices is wrong, the class must find a transformation that will bring the selected rock from the wrong sector to the right sector.

# 5.5 Global Climate Change

Global climate change may be one of the most challenging problems that humanity has ever faced. It will increasingly affect us in almost every aspects of our lives and will affect our children likely even more. While some of the Science is still developing given that we will be constantly faced with decisions regarding global climate change, this topic must be discussed with future teachers and young students.

But why should we care about global warming? For some warmer temperatures may be beneficial but for many the impacts will not be welcomed. These impacts include warmer temperatures on average for most of the earth, sea level rise, changes in precipitation patterns, etc. These changing conditions will affect how we live, how much fresh water is available in our region, what animals live around us, what type of agricultural products we cultivate in a given regions. While scientists almost unanimously agree on these points there is debate on exactly how the climate will change over specific parts of the earth, and how quickly it will change.

Facts are as often scientifically established first and then take some times to make their way to the public. For example while some still contest that we are experiencing a global temperature increase, the scientific community is virtually unanimous that we are experiencing a global warming, and that it is caused by the large increase of $CO_2$ and other greenhouse gases emitted as a result of human activities. To fully understand global climate change and make predictions one needs to understand how all the spheres of the earth interact with each other. Scientists put their knowledge of the functioning of the earth into large computer models called Global Climate Models. Scientists have been working on these models for many years but a lot more progress is still needed to make accurate predictions. Given that the science is constantly evolving on this topic, we need to stay in touch with scientists who have the best information and the experience and credentials to interpret it and give us guidance. For that reason the United Nations and the World Meteorological Organization created the Intergovernemental Panel on Climate Change or IPCC. The IPCC reviews the measurements, scientific reports and journal papers related to our climate and how it is changing. IPCC membership includes many distinguished scientists including Nobel prize winner, such as one of the scientists who helped us understand and control the ozone hole problem. The IPCC regularly updates on the status of the earth climate with the last major update dating back to 2007. Because IPCC looks at past data, their conclusions and recommendations lag a little behind the most recent developments. However this is the most reliable information for the state of our climate and its change.

**Definitions:**

*Weather:* Deals with the short term changes in the atmosphere, how warm/cold will it be tomorrow and during the next few days? Will it rain? Will it snow? Hail? The weather is ever changing and sometimes it can change in a mater of minutes or hours.

*Climate:* Deals with what is happening on average over a long period of time, decades, to our atmosphere. There are always warmer and colder years, rainy and dry years that follow each other. To say if the local climate is warming or cooling, if precipitation is increasing or decreasing, one needs to measure temperature or precipitation over many years and see if there is a trend.

# *(Web) Activity 5.12:*     **"Graphing Climate Change"**

This activity is based on the last broad assessment document from the International Panel on Climate Change (IPCC), "Climate Change 2007: The Physical Science Basis – Summary for Policymakers". The document can be found at the following website:

http://ipcc-wg1.ucar.edu/wg1/docs/WG1AR4_SPM_PlenaryApproved.pdf

You can also find the document by using the keywords 'IPCC' and '2007' on a search engine.

Based on the graphs and text of that document answer the questions and complete the graphs requested below.

1.  Based on figure SPM-3 how much has the earth averaged temperature increased between 1850 and 2005?

2.  Based on the same figure how much has the average sea level increased between 1870 and 2005?

3.  Based on figure SPM-4 which is warming up faster the oceans or the land?

4.  Download the following IPCC power point presentation:
    http://ipcc-wg1.ucar.edu/wg1/docs/Solomon_IPCCWG1.ppt

    Based on slide 16 what is approximately the ratio of locations on earth where the local climate is warming as compared to the number of locations where it is cooling? Find a location in the US that is cooling down.

5.  Reproduce in the graph below the change in carbon dioxide ($CO_2$) concentration in our atmosphere between 10,000 years ago and today. Use the unit of ppm or parts per million on the y-axis and years on the x-axis.

6. Based on slide 7 of the power point presentation, how much more $CO_2$ concentration do we presently have as compared to the maximum concentration over the past 650,000 years?

7. Give the name of other green house gases than $CO_2$ and their concentrations in our atmosphere in 2005. The process by which these gases contribute to global warming will be tackled in the next activity.

8. How much sea level rise would result from the melting of the Greenland glaciers (slide 28)?

9. Find on the web information on the population on earth and construct a bar graph below. You can use a search engine or use the following web site:

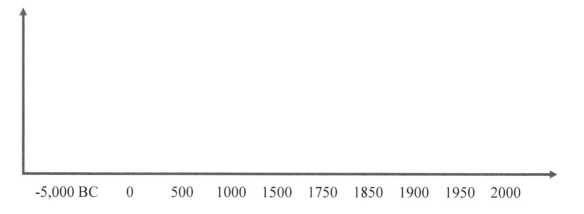

-5,000 BC    0    500    1000    1500    1750    1850    1900    1950    2000

10. Find a graph of the energy consumption versus standard of living index. Which among countries with a high standard of living use the most energy and which use the least? Discuss why you think there are such differences. You can use a search engine or use the following website: www.thewatt.com/node/168

Figure 5.10. A warming earth needing attention!

# Activity 5.13: "Solar Energy Accounting"

Everyone knows or at least should know how to balance a check book or an account. The same principle can apply to the balance of solar energy reaching the earth. We have a regular solar energy income, some of it is absorbed (taxed?) by the atmosphere, some of it is reflected back (bad checks?) to space. Once the solar rays hit the ground they can again be either reflected (automatic drafts?) or absorbed (savings?). But what does the earth do with this absorbed energy (savings?), well it reemits almost all this energy (spends it?). Actually the earth does not have a choice as the laws of Physics require that all warm objects emit light. This energy (savings) is emitted by the earth in the form of infrared rays. As these rays make their way towards space what can happen to them? They can make it to outer space, be absorbed by the atmosphere or be reflected back to earth. If they are absorbed by the atmosphere, the atmosphere in turn is heated and will reemit infrared rays back to earth or out to space.

In the tables below you will find information about how much light is reflected, absorbed, transmitted as it travels from space to earth and back. 100% corresponds to the total amount of solar energy, 341.3W/m$^2$, that reaches our earth. Start by adding percentages to the empty boxes.

| Focus on sunlight: Solar Radiation | | | | |
|---|---|---|---|---|
| Solar radiation coming towards earth | Solar radiation reflected by the atmosphere and clouds | Solar radiation absorbed by the atmosphere | Solar Radiation reaching the earth surface | Solar radiation directly reflected back towards the atmosphere |
| 100% | 23.2% | 22.9% | | 6.7% |

| Focus on the earth: Infrared radiation emitted by the earth or Earth Radiation | | | |
|---|---|---|---|
| Solar Radiation absorbed by the earth surface | Energy emitted by the atmosphere towards the earth | Total energy emitted by the earth towards the atmosphere | Approximate amount of energy absorbed by the earth (i.e. received but not reflected or reemitted) |
| 47.2% | 97.6% | 144.4% | |

If the number computed in the table above is positive, what does it mean for the earth?

| The Atmosphere: Atmosphere Radiation | | | | |
|---|---|---|---|---|
| Energy absorbed in the atmosphere from direct sunlight | Energy absorbed in the atmosphere coming from earth | Amount of radiation emitted by the atmosphere towards space | Amount of radiation emitted towards the earth | Solar radiation directly reflected back towards the atmosphere |
| 22.9% | 132.7% | 58.3% | 97.6% | 23.2% |

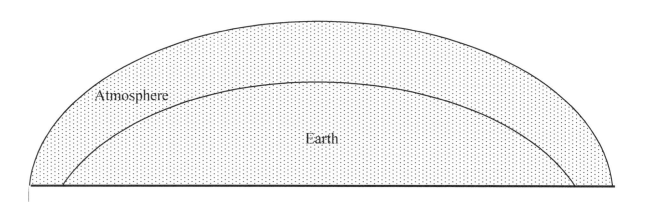

1. Take a yellow marker and draw a wide yellow ray coming from the sun and going towards the earth. Use the table in the previous page to split the ray once it reaches the atmosphere. There will be a smaller ray reflected by the atmosphere, another ray will be continuing through the atmosphere towards the earth surface. As it is going through the atmosphere this ray will be getting thinner because of absorption. Once it reaches the earth surface, part of the ray is absorbed (continue inside the earth) and part of the ray is reflected back.

2. Then take a red pen and use the data in the previous page to illustrate what happens to the radiation emitted by the earth (remember if energy is absorbed, the ray becomes thinner).

3. Finally take a green pen and follow the same procedure illustrate the energy emitted by the atmosphere back to the earth surface or towards space.

Give examples of events that can increase/decrease the amount of total radiation reaching the earth surface.


Give example of events that can increase/decrease the amount of radiation reflected by the earth surface.


What happens if gases that absorb efficiently infrared light are released to the atmosphere?

# 5.6 Beyond the Earth

For most of human history we knew very little about space beyond Earth. Early cultures from the Egyptians to the Greeks and from the Romans to the Mayans included celestial objects in their mythology and used them to keep track of natural cycles. However they did not know exactly what these objects were. Ancient people were able to distinguish two types of celestial objects based on their difference in motion across the sky: planets and stars. They even correctly concluded based on observations that stars are more distant than planets. But they wrongly assumed the position of the Earth placing it in the center of the Universe so stars, planets and the Sun circle around it.

The invention of the telescope in the 17th century gave us a closer look at planets. First telescopic observations helped us understand the structure and dynamics of the Solar System confirming that the Sun and not the Earth is the center of the Solar System. These observations also opened our mind to the fact that there are a lot more stars around us than we initially thought. The aid of telescopes opened our eyes to the vast distances between the stars and helped us form our understanding of the Milky Way Galaxy.

The next step taken was literally out into space: Sputnik was the first artificial satellite launched in 1957. Since then, satellites and other robotic instruments have been our main vanguard. They gave us images of the dark side of the Moon and close-up pictures of all the planets of the Solar System. Placing a telescope on a satellite in orbit has allowed us to see much more of the Universe and recently we have been able to confirm that other stars also have planets orbiting around them. Since 1996 Mars Rovers have been rolling across the surface of Mars using probes to analyze its soil. The robotic New Horizon spacecraft is currently charting its course to Pluto and other objects of the Kuiper Belt. We now recognize that Space is no longer a strange and dark place beyond Earth and we see ourselves as part of the highly complex Universe. There is much more left to explore and the coming years will give us many more discoveries. Mankind has already walked on the surface of the Moon and who knows maybe in the future there will be a second planet with a permanent human presence.

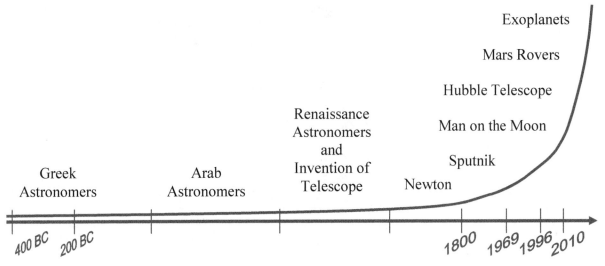

Figure 5.11. Timeline of our knowledge of space

157

# Chart 5.11: "Celestial Hierarchy"

**UNIVERSE**
Estimated to be about 13.7 billion years old
Has been expanding ever since (red shift of galaxies)

**CLUSTER OF GALAXIES**
From small clusters as Local Group of about 30galaxies (including our galaxy) to large clusters like the neighboring Virgo Cluster -1300 galaxies

**GALAXY**
Our Milky Way has estimated 200-400 billion stars
Every star you see in the sky with unaided eyes is part of the Milky Way Galaxy

**STAR**
Produces energy by fusion of light elements
Our star, Sun, is part of the Milky Way Galaxy

**PLANETARY SYSTEM**
Our Solar System has 8 planets
and belongs to the star SUN

**PLANET**
Does not emit but reflects star light
We live on planet Earth, the 3rd planet of Solar System

The Solar System was borne about 4.6 billion years ago from gas and dust of previously existing stars. These remnants of previous stars are called a **nebula**. Our nebula was eventually pulled together by gravity into a flat spinning disk from which the Sun and the planets were formed. All of the planets orbit the Sun in the same direction and do not cross each other's orbits; planetary orbits are defined to the equatorial plane of the Sun and are spaced out progressively with the distance from the Sun. Besides planets, the Solar System has two groups of small objects: asteroids located between the orbits of Mars and Jupiter and KBO's or Kuiper Belt Objects located beyond the orbit of Neptune. KBO are the remains from the formation of the solar system swiped away by the first solar wind. Some have orbits that make these icy-rocky mixtures visit the inner solar system periodically. We call them comets. When a comet passes through the warm part of the inner solar system it thaws out and forms a gas cloud around it. This cloud, swiped by the solar wind, makes a tail that we can see from the Earth. After a comet passes, small rocks lost from the comet as the result of thawing are left behind and continue to orbit the Sun as meteoroids. When Earth is passing through a stream of such meteoroids they burn in Earth's atmosphere and we see "shooting stars" or a meteor shower.

Recently the International Astronomical Union defined another group of Solar System objects: the dwarf planets. A few KBOs including Pluto and a largest asteroid Ceres now fall under this category.

## 2009 Solar System Consensus

- **1 star**
- **8 planets**
- **63 major natural satellites of planets**
- **Over thousand of Kuiper Belt Objects (including Pluto and other "dwarf planets")**
- **A few hundred thousands Asteroids (including "dwarf planet" Ceres)**
- **Comets and Meteoroids**

Our star Sun is located almost at the edge of the galaxy which is why we do not see very many stars around us even out of city lights. Away from lights and during dry nights one can see in the night sky a long whitish band called Milky Way; it is the side view of our galactic disk with a higher concentration of star.

Our planet Earth is moving along an elliptical (but close to circular) orbit around the Sun and is the third planet from the Sun. The Earth rotates on its axis once every 24 hours. As the result of this rotation half of the Earth facing the sun has daytime while the other half has nighttime. The rotational axis of Earth is tilted to the orbital plane. While the tilt does not change as the earth revolves around the sun in one year, it is still responsible for the seasonal changes on Earth. One of the hemispheres is always tilted towards the Sun. The sunlight hits the surface of this hemisphere at a steeper angle than the surface of the other hemisphere that is tilted away from the Sun. The steeper the angle the more amount of light received by the hemisphere and it experiences summer. The shallower the angle the less amount of light received by the hemisphere and it experiences winter. During the northern hemisphere summer time, the southern hemisphere has winter and vice and versa. The regions between the tropics do not have large variation in the angle of sunlight, and therefore have no seasons.

# Chart 5.12:         "Family Tree"

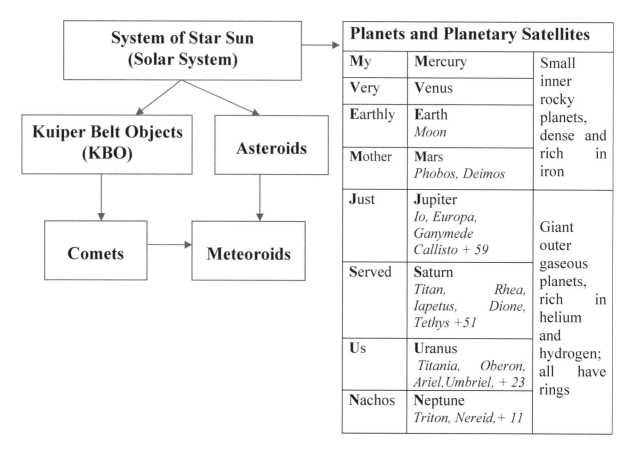

**System of Star Sun (Solar System)**

**Kuiper Belt Objects (KBO)**

**Asteroids**

**Comets**

**Meteoroids**

| Planets and Planetary Satellites | | |
|---|---|---|
| My | Mercury | Small inner rocky planets, dense and rich in iron |
| Very | Venus | |
| Earthly | Earth *Moon* | |
| Mother | Mars *Phobos, Deimos* | |
| Just | Jupiter *Io, Europa, Ganymede Callisto + 59* | Giant outer gaseous planets, rich in helium and hydrogen; all have rings |
| Served | Saturn *Titan,    Rhea, Iapetus,    Dione, Tethys +51* | |
| Us | Uranus *Titania,   Oberon, Ariel, Umbriel, + 23* | |
| Nachos | Neptune *Triton, Nereid, + 11* | |

Figure 5.12. The sun and the solar system planets in order with size approximately up to scale.

# Chart 5.13: "Planets of the Solar System"

| Planet | Famous for... |
|--------|---------------|
| **MERCURY** | Famous for its position - closest to Sun - fastest small planet. Very small - weak gravity. Weak gravity- can not hold an atmosphere. |
| **VENUS** | Famous for its extreme environment. Heavy carbon dioxide atmosphere causes greenhouse effect results in enormous surface pressure and temperature. Called "morning/evening" star because of time when it is visible |
| **EARTH** | Famous for its ability to sustain life. Midsize volcanically active rocky planet with liquid water. |
| **MARS** | Famous for its red color caused by iron oxide dust covering the surface. Little more than half Earth size – holds little atmosphere. Little atmosphere -very low pressure. Low atmospheric pressure -water can not exist as liquid. Water exists as ice in polar caps, as permafrost or an atmospheric vapor. |
| **JUPITER** | Famous for its size. The largest of all the planets. Has very strong magnetic field. Does not have a hard surface, more than half is an ocean of liquid hydrogen. Visible stripes and ovals on the surface are jet streams and hurricanes. |
| **SATURN** | Famous for its rings visible even through binoculars. Rings are chunks of frozen water and rocks. Has density lower than water. |
| **URANUS** | Famous for its tilt. It lies on its side, probably knocked off by ancient collision. |
| **NEPTUNE** | Famous for its discovery not by an observation but a calculation. Bright blue color is due to methane atmosphere |

**How did all it start?** The main observation that a theory of the universe must include is that all groups of galaxies are moving away from each other (observation of red shift of star lights). In 1929 Monsignor Lemaître, a Belgian Roman Catholic priest and professor of Physics and Astronomy came up with the theory that it all started from a single point and that universe evolved from there: *the Big Bang*. For many years the theory was not popular. It took another observation to make the theory our dominant one. In 1964 Arno Penzias and Robert Wilson measured microwave radiations coming from everywhere in the skies consistent with the theory of the big bang. The Big Bang theory is by far our best scientific explanation. There is presently no significant scientific theory as to where the initial energy came from what got our universe stated. Science being ultimately based on observations and experiments, we may never have a scientific theory to explain the start of the universe. As illustrated in Figure 5.x and table 5.x it takes a while for light to reach us. When we look at light from a start we are looking in the past of that star. If the start is very far away it may not exist anymore when we observe its light. As nothing goes faster than the speed of light we cannot look at light that we emitted ourselves. In other words we are not able to observe our own past.

**Distances in the universe**: Because distances are so large, instead of using kilometers or miles, one uses the distance traveled by light. In 1 year a ray of light in vacuum travels, $9.5 \times 10^{12}$ km. In the table below you will find distances between our earth and other universe landmarks.

Table 5.x  Travel time to locations in space at the speed of light and the speed of a Boeing 757.

| Celestial Objects | Speed of Light | Speed of 757 |
|---|---|---|
| Earth – Moon | 1.3 seconds | 21 days |
| Earth – Sun | 8 minutes | 21 years |
| Sun – nearest other star: Proxima Centauri | 4.3 years | 6 million years |
| Sun – Center of the Milky Way Galaxy | 25,000 to 30,000 years | 34 to 41 billion years |
| Diameter of the Milky Way Galaxy | 200,000 years | 275 billion years |
| Nearest other galaxy: Andromeda (can only be seen through a telescope) | 2 million light years | 2.75 trillion years |

 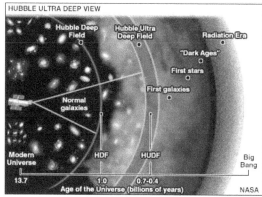

Firgure 5.13. (a) The Hubble telescope (b) relationship between distance of observed light and time when the light was emitted. Graphics courtesy of NASA.

The earth has one natural satellite, Moon. We always see the same side of the moon from earth as the rotation of the moon around itself is synchronized with its rotation around the earth.

## *Activity 5.14:* "Phases of the Moon"

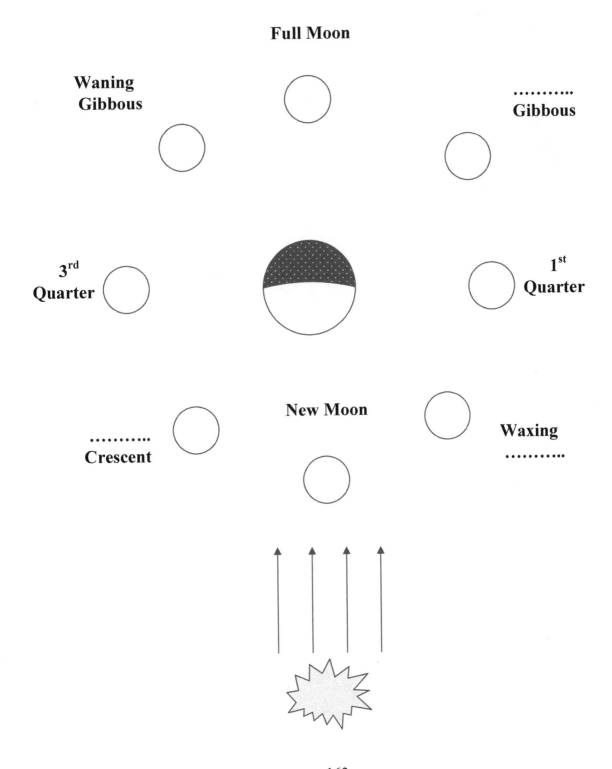

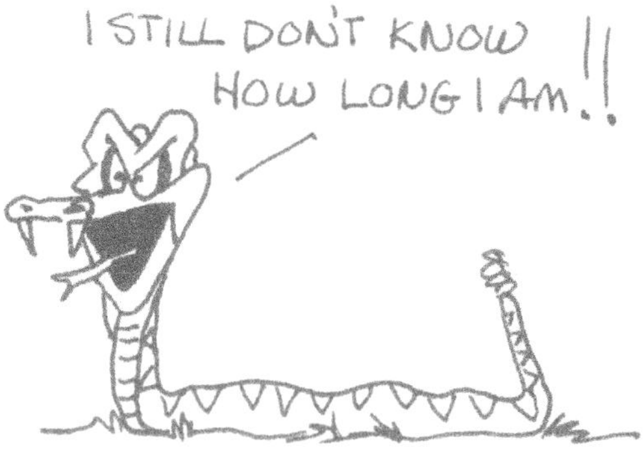


DON'T PANIC....

CPSIA information can be obtained
at www.ICGtesting.com
Printed in the USA
LVHW061938091118
596602LV00002B/4/P